W9-AGJ-196

THE PICTORIAL HISTORY OF

NASA

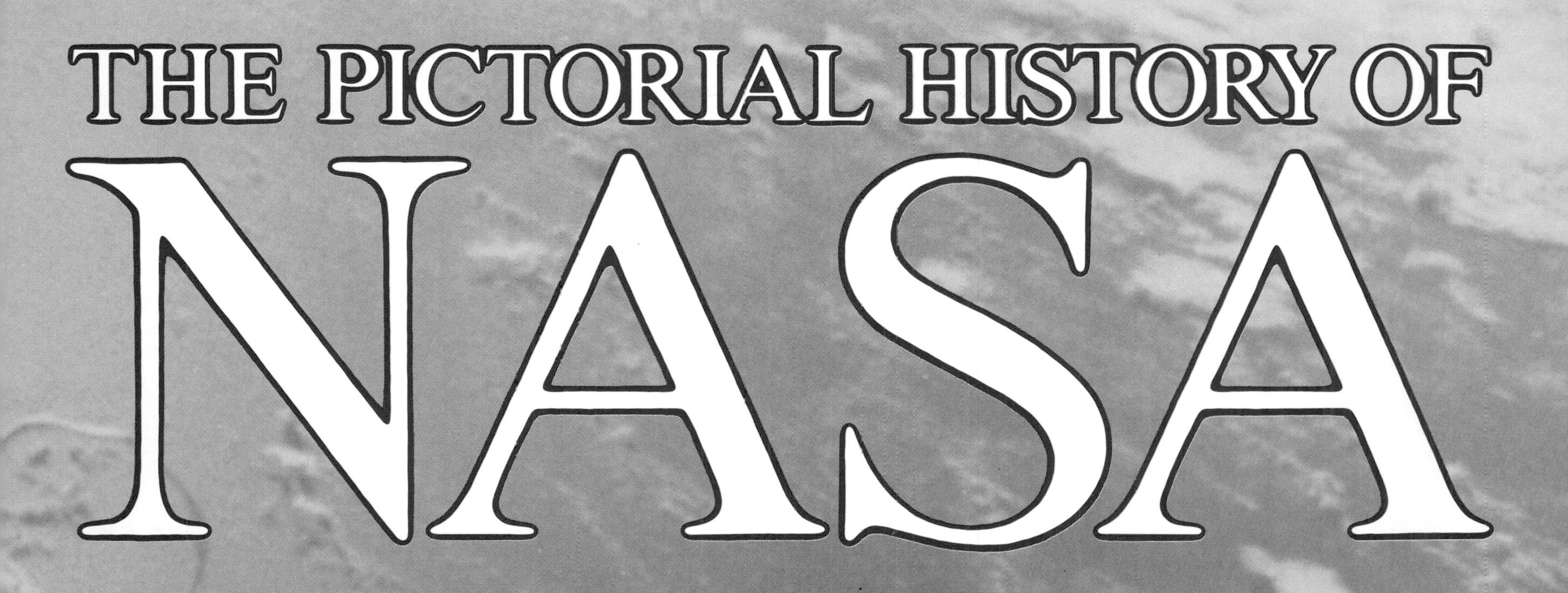

THE PICTORIAL HISTORY OF NASA

Edited by
BILL YENNE

GALLERY BOOKS
An imprint of W.H. Smith Publishers Inc.
112 Madison Avenue
New York, New York 10016

Published by Gallery Books
A Division of W H Smith Publishers Inc.
112 Madison Avenue
New York, New York 10016

Produced by
Brompton Books Corp.
15 Sherwood Place
Greenwich, CT 06830

ISBN 0-8317-6901-7

Printed in Hong Kong

10 9 8 7 6 5 4 3 2 1

Picture Credits

All photos were supplied through the generosity of the National Aeronautics and Space Administration with the following exceptions:

Boeing Aerospace Company: 1, 98–99 (right)
Department of Defense: 32, 178 (top)
Hughes Aircraft Company: 53, 81, 88-89 (all), 93, 125 (bottom)
Lockheed Corporation : 171, 172-173 (right), 178 (middle)
McDonnell Douglas Corporation: 108, 115 (bottom), 117, 119 (top), 120–121
Rockwell International Corporation: 16, 98 (left), 100 (bottom), 146 (all), 176,178 (bottom)
Smithsonian Institution National Air & Space Museum: 64 (top)
US Air Force: 36 (right)
©Bill Yenne: 96-97 (bottom)

Acknowledgements

We would also like to extend a special thanks to Mike Gentry at NASA's Johnson Space Center in Houston for his invaluable assistance in producing this work.

Designed by Tom Debolski
Edited by Bill Yenne
Captioned by Timothy Jacobs

Page 1: **A massive Saturn 5 booster rides its transport en route to the launch pad. The Saturn 5 was the most powerful booster ever built by the US, and was designed expressly for the Apollo launches.**

Pages 2—3: **This orbital photo captures Astronaut Dale A Gardner on a 14 November 1984 spacewalk outside of Shuttle Orbiter *Discovery*, during Space Shuttle mission 51-A.**

These pages: **The then-brand new Orbital Vehicle *Atlantis* is shown in transit in this photo. Atlantis had its maiden voyage with Space Shuttle Mission 51-J, on 3 October 1985.**

CONTENTS

INTRODUCTION

by Bill Yenne

The National Aeronautics & Space Administration (NASA) was officially chartered on 1 October 1958, to 'provide for research into problems within and outside the earth's atmosphere, and for other purposes.' NASA was the world's first civilian space agency and the first entity anywhere to have the exploration of space as the cornerstone of its *raison d'être*.

Prior to the creation of NASA, American efforts directed toward space exploration had consisted primarily of the competing activities of the US Army and the US Navy. The United States had hoped to put its first satellite into orbit around the earth in 1957, during the International Geophysical Year (IGY). The US Navy's Vanguard program, begun in 1955, was seen as the most likely to succeed, but nothing could have been farther from the truth. Both delays and launch vehicle failures made Vanguard almost ridiculous. The Navy's embarrassment was compounded when the Soviet Union launched Sputnik 1, the world's first satellite, on 4 October 1957. The Army finally managed to orbit America's first satellite, Explorer 1, on 31 January 1958, but by that time the Russians had orbited a *second* satellite—and Sputnik 2 successfully carried a dog into space.

It immediately became clear that the United States could not run a successful space program against a background of intramural competition. The Eisenhower administration decided in 1958 that the United States should make a serious effort to become a world leader (if not *the* world leader) in space exploration, and that this activity should all be consolidated under a single non-military agency. The model for the new space agency would be The National Advisory Committee for Aeronautics (NACA), a similar entity created in 1911 to deal with the emerging science of aeronautics. Indeed, because the new space agency would *also* be chartered to study aeronautics, it was decided to absorb NACA *within* the new agency. Even the initials of the new agency would be reminiscent of NACA.

The new NASA, as well as a National Aeronautics and Space Council presided over by the President of the United States, were created by the National Aeronautics & Space Act of 1958, which was passed by Congress on 29 July. The Act stated in part:

'The Congress hereby declares that it is the policy of the United States that activities in space should be devoted to peaceful purposes for the benefit of all mankind.

'The Congress declares that the general welfare and security of the United States require that adequate provision be made for aeronautical and space activities. The Congress further declares that such activities shall be the responsibility of, and shall be directed by, a civilian agency exercising control over aeronautical and space activitities sponsored by the United States, except that activities peculiar to or primarily associated with the development of weapons systems, military operations, or the defense of the United States (including the research and development necessary to make effective provision for the defense of the United States) shall be the responsibility of, and shall be directed by, the Department of Defense; and that determination as to which such agency has responsibility for and direction of any such activity shall be made by the President.

'The aeronautical and space activities of the United States shall be conducted so as to contribute materially to one or more of the following objectives:

(1) The expansion of human knowledge of phenomena in the atmosphere and space;

(2) The improvement of the usefulness, performance, speed, safety, and efficiency of aeronautical and space vehicles;

(3) The development and operation of vehicles capable of carrying instruments, equipment, supplies, and living organisms through space;

(4) The establishment of long-range studies of the potential benefits to be gained from, the opportunities for, and the problems involved in the

At right: **Vanguard TV-1 on the pad. This was one of many attempts to launch a Vanguard—few were successful.** ***Above:*** **Vanguard SV-2's Satellite is here shown previous to its unsuccessful mission of 26 June 1958.** ***Below:*** **The Vanguard SLV-5's magnetic field and atmospheric physics satellite, destined for launch failure.**

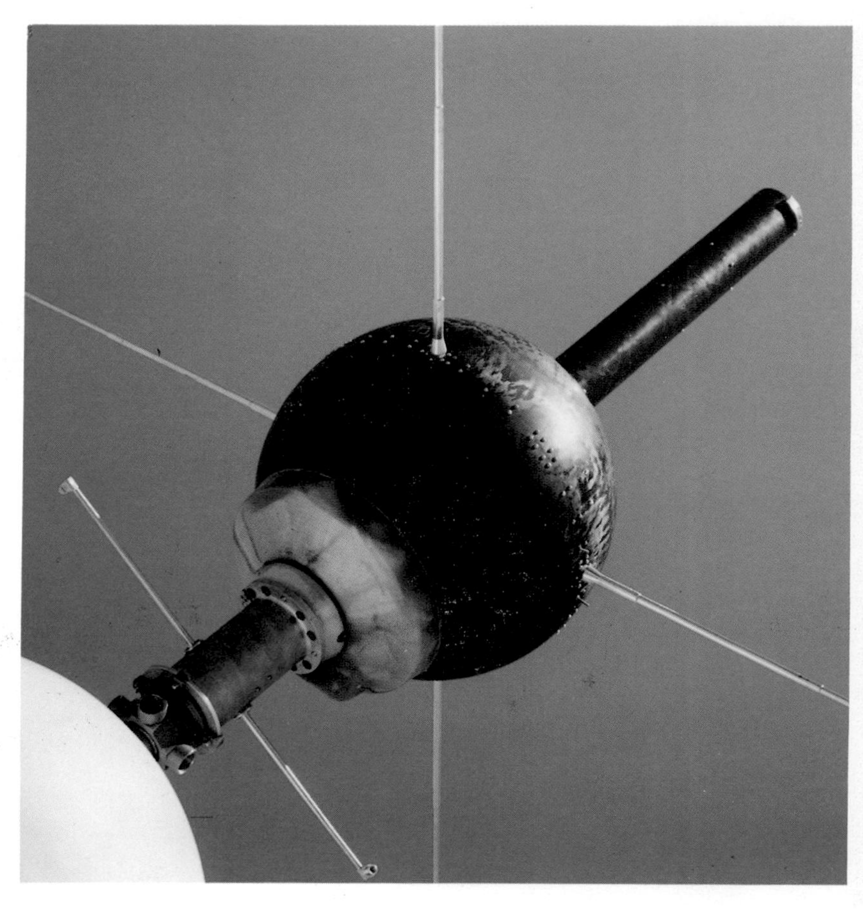

IV

utilization or aeronautical and space activities for peaceful and scientific purposes;

(5) The preservation of the role of the United States as a leader in aeronautical and space science and technology and in the application thereof to the conduct of peaceful activities within and outside the atmosphere;

(6) The making available to agencies directly concerned with national defense of discoveries that have military value or significance, and the furnishing by such agencies, to the civilian agency established to direct and control nonmilitary aeronautical and space activities, of information as to discoveries which have value or significance to that agency;

(7) Cooperation by the United States with other nations and groups of nations in work done pursuant to this Act and in the peaceful application of the results thereof; and

(8) The most effective utilization of the scientific and engineering resources of the United States, with close cooperation among all interested agencies of the United States in order to avoid unneccessary duplication of effort, facilities, and equipment.'

In the more than thirty years since it was created, NASA has done all this and more. The agency has developed and deployed an archipelago of weather and communications satellites that have changed the way we live and the way we do business. In terms of space science, NASA has sent spacecraft to the distant reaches of our Solar System and has landed a robot weather station on the surface of Mars.

NASA has managed all of America's manned spaceflight programs. When President John F Kennedy committed the nation to putting an American on the moon by 1969 and returning him safely to Earth, NASA answered and met the challenge. In fact, NASA sent and safely returned *four* astronauts by the end of 1969 and another eight by the end of 1972. NASA built and deployed America's only space station — Skylab — which was the *largest* space station ever launched.

Finally, in 1981, NASA first successfully launched its 'Space Shuttle,' a system which contained the revolutionary airplane-like manned Orbiting Vehicle capable of 'flying' into the Earth's atmosphere and returning to space an indeterminate number of times.

With thirty years of activity behind it, NASA has Voyager spacecraft flying amid the outer planets and Shuttles carrying Americans routinely into orbit. NASA's first three decades have been marked by both triumph and tragedy, but the emphasis is on the former and all eyes are on tomorrow. This book is the history of those thirty thrilling years, but please note that these years are just the beginning, and the best is yet to come!

Below: **A Vanguard launch. Success finally came to Vanguard on 17 March 1958.** ***At right:*** **Almost 30 years later, on 6 November 1985, the Shuttle Transportation System completes yet another of a long series of successful flights with the landing of** ***Challenger*** **STS-61A, but tragedy struck** ***Challenger*** **on 27 January 1986.**

NASA
United States

NASA Installations

by Leon Perry

Since NASA's establishment in October 1958, its network of centers and facilities has spread across the United States. It is at these field installations that NASA conducts its many scientific programs, ranging from aerodynamic research to make commercial aviation safer here on earth, to sending spacecraft to the reaches of space in a search for life on other planets.

NASA HEADQUARTERS
Washington, DC

Headquarters manages the space flight centers, research centers and other NASA installations. Planning, direction and management of NASA's research and development programs are the responsibility of individual program offices which report to, and are directed by, Headquarters officials.

Headquarters responsibilities include the determination of projects and programs, establishment of management policies, procedures and performance criteria, and review and analysis of all phases of the aerospace program.

AMES RESEARCH CENTER
Mountain View, California

The NASA Ames Research Center (ARC) is located at the south end of San Francisco bay near Mountain View, California. San Francisco is 35 miles to the northwest and San Jose is ten miles to the southeast. The US Naval Air Station, Moffett Field, California, is contiguous to the south and east.

The programs of the center are directed towards research and development of technology in the fields of aeronautics, space science, life science and spacecraft technology.

Ames operates specially outfitted aircraft — in effect, airborne laboratories to serve as flying instrument platforms for the use of scientists from all over the world in studies of both space and earth.

Ames has management responsibility for the Pioneer series of spacecraft; six of which are in solar orbit and two others which will become the first man-made objects to escape the solar system after providing closeup pictures of Jupiter and Saturn. Yet another Pioneer is still returning data from the planet Venus.

In life science laboratories, Ames scientists study the origin of life and the relationship between man and aircraft and provide medical criteria for allowing man on space vehicles.

In spacecraft technology, Ames supports NASA's Space Shuttle program by providing research on heat protection systems and wind tunnel investigations of the stability and heating of the various configurations and/or modifications.

In cooperation with Federal, state and local agencies, Ames conducts pilot programs and prototype investigations of applications of space technology to earth-bound problems.

Ames Research Center also exercises management control for the Dryden Flight Research Facility, Edwards, California.

HUGH L DRYDEN FLIGHT RESEARCH FACILITY
Edwards AFB, California

The Dryden Flight Research Facility, a Directorate of NASA's Ames Research Center, is located at Edwards AFB, California, on the edge of the Mojave Desert, approximately 80 miles north of Los Angeles.

It was established as the major NASA facility for high speed flight tests. The primary research tools for conducting the programs and missions of the facility are its aircraft. They range from Century Series fighters to advanced supersonic and hypersonic aircraft and aerospace flight research vehicles

such as wingless lifting bodies. There are also special ground-based facilities, including a Flight Test Range with a fully instrumented tracking station; a high temperature loads calibration laboratory; and a remotely piloted research vehicle (RPRV) facility.

Dryden's mission includes research and test activities on problems of aircraft and flight research vehicle takeoff and landing, low speed flight, supersonic and hypersonic flight, and flight vehicle reentry to verify predicted vehicle and flight characteristics and to identify unexpected problems in actual flight and to correlate theoretical or wind tunnel research studies. The facility also conducted studies into terminal area operations of the Space Shuttle vehicle and flight investigations involving the flight tests of the Space Shuttle.

ROBERT H GODDARD SPACE FLIGHT CENTER
Greenbelt, Maryland

The Goddard Space Flight Center (GSFC) is located ten miles northeast of Washington, DC. Goddard consists of facilities in Greenbelt, Maryland, Wallops Island, Virginia, Goddard Institute for Space Studies in New York City, and 16 tracking stations around the world.

The Goddard Space Flight Center conducts and is responsible for automated spacecraft and sounding rocket experiments in support of basic and applied research. Satellite and sounding rocket projects at Goddard provide data about the earth's environment, sun/earth relationships and the universe. These projects also advance technology in such areas as communications, meteorology, navigation and the detection and monitoring of our natural resources.

Goddard is the home of the National Space Science Data Center, the central repository of data collected from space flight experiments.

Goddard personnel are situated around the globe as part of the Space Tracking Data Network (STDN) team and at facilities of the NASA Communications Network which links these networks together.

The center serves as project manager for the Delta launch vehicle and has also been assigned a lead role in the management of the international Search and Rescue Satellite Aided Tracking (SARSAT) project.

Much of the center's theoretical research is conducted at the Goddard Institute for Space Studies in New York City.

JET PROPULSION LABORATORY
Pasadena, California

NASA's Jet Propulsion Laboratory, (JPL) is located in Pasadena, California, approximately 20 miles northeast of Los Angeles.

Jet Propulsion Laboratory is a government-owned facility that is staffed and managed by the California Institute of Technology. At Pasadena, JPL occupies 177 acres of land, of which 155 are owned by NASA and 22 acres are leased. JPL operates under a NASA contract, which is administered by the NASA Pasadena office. In addition to the Pasadena site, JPL operates the Deep Space Communications Complex, a station of the worldwide Deep Space Network (DSN) located at Goldstone, California, on 40,000 acres of land occupied under permit from the Army.

The laboratory is engaged in activities associated with deep space automated scientific missions, tracking, data acquisition, data reduction and analysis required by deep space flight, advanced solid propellant and liquid propellant spacecraft engines, advanced spacecraft guidance and control systems, and integration of advanced propulsion systems into spacecraft.

The laboratory designs and tests flight systems, including complete spacecraft, and also provides technical direction to contractor organizations.

NASA's Ames Research Center in Mountain View, California has the world's largest wind tunnel—one of the air intakes to which is shown at *below left*. NASA's Dryden Flight Research Center was the site of the scene *below*, which shows the Shuttle Orbiter test vehicle being mated to its 'mother ship,' a modified Boeing 747.

JPL operates the worldwide deep space tracking and data aquisition network, the DSN and maintains a substantial program to support present and future NASA flight projects and to increase capabilities of the laboratory.

LYNDON B JOHNSON SPACE CENTER

Houston, Texas

Johnson Space Center is located on NASA Highway 1, adjacent to Clear Lake, two miles east of the town of Webster, and approximately 20 miles southeast of downtown Houston. Additional JSC facilities are located at nearby Ellington Air Force Base, which is approximately seven miles to the north of the center.

Johnson Space Center was established in September 1961 as NASA's primary center for design, development and manufacture of manned spacecraft; selection and training of space flight crews, for ground control of manned flights and many of the medical, engineering and scientific experiments carried aboard the flights. JSC is the lead NASA center for management of the Space Shuttle.

One of the center's best known facilities is the Mission Control Center from which manned flights, starting with Gemini 4, through Apollo and Skylab series, and continuing into the current missions of the Space Shuttle, are monitored.

JSC is also responsible for direction of operations at the White Sands Test Facility (WSTF), located on the western edge of the US Army White Sands Missile Range at Las Cruces, New Mexico. WSTF supports the Space Shuttle propulsion system, power system and materials testing. The facility also served as the alternate landing site for the second test flight of the orbiter Columbia.

JOHN F KENNEDY SPACE CENTER

Cape Canaveral, Florida

Kennedy Space Center (KSC) is located on the east coast of Florida, 150 miles south of Jacksonville and approximately 50 miles east of Orlando. It is immediately north and west of Cape Canaveral. The Kennedy Space Center is about 34 miles long and varies in width from five to ten miles. The total land and water area occupied by the installation is 140,393 acres. This area, with adjoining water bodies, provides sufficient space to afford adequate safety to the surrounding civilian community for planned vehicle launchings.

Kennedy Space Center serves as the primary center within NASA for the test, checkout and launch of space vehicles. This presently includes launch of manned and unmanned vehicles at the Kennedy Space Center and Air Force Eastern Space and Missile Center in Florida, and the Air Force Western Space and Missile Center at Vandenberg Air Force Base in California. The center is now concentrating on the assembly, checkout and launch of Space Shuttle vehicles and their payloads, landing operations and the turnaround of Space Shuttle orbiters between missions, as well as research and operational unmanned launches.

Kennedy Space Center is also responsible for the operation of the KSC Space Transportation System (STS) Resident Office, located at the Vandenberg AFB in Santa Barbara County, on the California central coast. The KSC STS Resident Office provides or arranges host base support for all NASA elements at Vandenberg AFB and range support for all STS and

NASA's Jet Propulsion Laboratory manages such global Deep Space Network stations as the one *below*, in Australia. *Below right:* Launch Pad 39A at Kennedy Space Center—the launch site for the manned Apollo lunar missions, and all Space Shuttle and Skylab launches. *Above right:* Mission Control at Johnson Space Center.

These pages: The large, square building in this photograph is the Space Shuttle assembly building at Kennedy Space Center, where the Shuttle is mated to its solid booster and huge LOX fuel tank. After its stint in this building, the launch-ready Shuttle is trucked over to Pad 39A, which is not shown here.

Kennedy Space Center Deployable Payload project requirements. The Resident Office supports the Air Force in the design, construction, and activation of the Space Shuttle Vandenberg Launch and Landing site; provides support for all NASA Deployable Payload Operations; and assists the KSC Cargo Projects Office in planning for all STS cargo operations at Vandenberg.

LANGLEY RESEARCH CENTER
Hampton, Virginia

The Langley Research Center (LaRC), located in Hampton, Virginia, is approximately 100 miles south of Washington, DC. It is situated in the Tidewater area of Hampton Roads, between Norfolk and Williamsburg, Virginia. Langley's primary mission is the research and development of advanced concepts and technology for future aircraft and spacecraft systems, with particular emphasis on environmental effects, performance, range, safety and economy.

The aeronautical research program is aimed at identifying and pursuing basic and applied research opportunities offering the greatest potential for increases in performance, efficiency and capability. Included in the research laboratories are a variety of wind tunnels covering the entire Mach-number and Reynolds-number range.

THE LEWIS RESEARCH CENTER
Cleveland, Ohio

Lewis Center, (LRC) is located on the west side of Cleveland's Hopkins Airport in Cuyahoga County, Ohio.

Lewis is the primary NASA center for research and technology development in aircraft propulsion, space propulsion, space power and satellite communication. The center conducts research on power systems for converting chemical and solar energy into electricity.

Major research tools at the center are designed to simulate flight conditions and range from atmospheric wind tunnels to large space environmental facilities.

Lewis also manages the Centaur launch vehicle, a second-stage vehicle used with the Atlas first stage, and to be used in a modified configuration with the Shuttle.

GEORGE C MARSHALL SPACE FLIGHT CENTER
Huntsville, Alabama

Marshall Space Flight Center (MSFC), is located on Redstone Arsenal, just outside Huntsville, Alabama, and is easily accessible from Interstate 65 and US 231 with deep-water access to the Tennessee river.

Located on 1840 acres of land, Marshall is one of the nation's leading pioneering space centers. It was established in 1960 by a team of former Army rocket experts headed by Dr Werner Von Braun.

The center provides project management as well as scientific and engineering support for many of NASA's prime space programs and scientific endeavors. It has a wide spectrum of technical facilities and laboratories.

Originally NASA's primary propulsion development center, MSFC has diversified into a center for the development of payloads and space science activities.

The Marshall center is responsible for managing the development of the Space Shuttle main engines, solid rocket boosters and external propellant tank. It is also responsible the Space Telescope, the Spacelab orbital research facility and for many other key research and development programs.

Marshall also operates the Michoud Assembly facility in New Orleans, Louisiana, and the Slidell Computer Complex (SCC) at Slidell, Louisiana.

NATIONAL SPACE TECHNOLOGY LABORATORIES
Bay St Louis, Mississippi

The National Space Technology Laboratories (NSTL) is located in Hancock County, near Bay St Louis, Mississippi, on the east Pearl River. The installation is situated midway between New Orleans, Louisiana, and the Mississippi Gulf Coast.

Formerly designated the Mississippi Test Facility, NSTL was given full field installation status by NASA in 1974, because of its capabilities in space applications and earth resources activities. The Saturn rocket test stands have been modified for Marshall Space Flight Center main engine testing and for orbiter main propulsion testing for the Space Shuttle program.

The mission of NSTL is support of Space Shuttle main engine and main orbiter propulsion system testing. Static test firing is conducted on the same huge test towers used from 1965 to 1970 to captive-fire all first and second stages of the Saturn 5 used in the Apollo manned lunar landing and Skylab programs. Shuttle main engine testing has been under way at NSTL since 1975.

WALLOPS FLIGHT FACILITY
Wallops Island, Virginia

The Wallops Flight Facility (WFF), a part of the Goddard Space Flight Center, is located in Virginia on the Atlantic Coast, Delmarva Peninsula. It is approximately 40 miles southeast of Salisbury, Maryland, and 72 miles north of the Chesapeake Bay Bridge Tunnel. The facility includes three separate areas on the Atlantic Coast of Virginia's Eastern Shore: the main base, the Wallops Island launching site and the mainland site, plus 1140 acres of marshland. Wallops Island is about seven miles southeast of the main base and is five miles long and a half mile wide at its widest.

Wallops is responsible for managing NASA's Suborbital Sounding Rocket Projects from mission and flight planning to landing and recovery, including: payload and payload carrier design, development, fabrication, and testing; experiment management support; launch operations; and tracking and data acquisition.

The National Space Technology Laboratories support Space Shuttle main engine and main orbiter propulsion testing. *Below* is shown a Shuttle Transportation System main engine test. *At right* is shown the Wallops Flight Facility's long-range radar antennae, located on that facility's mainland site. *Below right* is an aerial view of the Wallops Flight Facility main base storage buildings and launch sites.

SPACE SCIENCE

by Victor Seigel

NASA's space science programs have contributed significantly to a new golden age of discovery. They have substantially advanced the frontiers of knowledge about our home planet, the relationships of sun and earth and celestial phenomena.

NASA satellites discovered the existence of the Van Allen Radiation Region around earth. They demonstrated that earth's magnetic field is not shaped like iron filings around a bar magnet but like a vast cosmic teardrop. It is literally blown into this shape by the solar wind, a hot electrified gas constantly speeding out from the sun.

Satellites also contributed to understanding of interaction of solar activity with the magnetic field to produce periodic radio black-outs, trip circuit breakers of electric transformers, cause magnetic compasses to become erratic, and generate auroras.

NASA satellites demonstrated that earth's tenuous upper atmosphere is not as stable and quiescent as previously believed. It swells by day and contracts at night. Its volume and density wax and wane with such solar events as solar flares, the 11-year solar cycle and the 27-day solar-rotation period.

NASA spacecraft confirmed existence of the solar wind. They discovered that the solar wind streams outward along the sun's magnetic field lines. Analyzing magnetic field and solar wind data from spacecraft near and far, scientists have redrawn the picture of interplanetary space. The solar wind and solar magnetic field are now visualized as forming a vast heliosphere encompassing space billions of kilometers outward from the sun.

Satellites have dramatically altered our conception of the universe. Our atmosphere blocks most of the electromagnetic radiations that can tell us about the nature of celestial objects. Satellite observatories viewing the heavens from above our appreciable atmosphere open a window on the universe.

Astronomers had theorized that only a small fraction of the electromagnetic radiation emitted by stars consisted of X-rays. (X-rays are generated by high energy processes; a sparse emission of X-rays would indicate that our universe was comparatively peaceful and slowly evolving.) However, NASA satellites revealed a skyful of X-ray sources. This revolutionized the concept of the universe to one whose dynamics and evolution are governed by dramatic and enormously powerful processes. Our study of the universe draws us toward the answers to fundamental questions about the very nature of matter, life and the destiny of the stars.

Arguably, the most dramatic discoveries have resulted from observations of other planets and, if present, their satellites.The observations accentuated the uniqueness of our planet, its resources and its environment. It is a place where a variety of suitable conditions combined to create and sustain life.

Our nearest neighbor, the moon, is a radically different world. NASA's Apollo expeditions and unmanned spacecraft, as well as telescope observations, show it to be pockmarked with huge meteorite craters partially filled with basalts from ancient lava flows. Its surface is wracked with excessive heat by day and cold by night. It is bombarded by solar radiation because it has neither an intrinsic atmosphere nor magnetic field to ward off the radiation. It has no signs of life nor even evidence that life processes have begun.

In times past, some people speculated that Venus was a twin of earth. Venus' density, gravity and size are nearly the same as earth's. The white clouds enveloping the planet were likened to water-based clouds of earth.

Spacecraft gave starkly different views. The beautiful white clouds are composed primarily of corrosive sulphuric acid droplets. Venus' atmosphere, 97 percent of which is carbon dioxide, has a crushing surface pressure about 100 times earth's at sea level. Its surface temperature is 900° F, hot enough to melt lead or zinc. Water has disappeared from the hot planet. Spacecraft confirmed that Venus' oven-like surface temperature is

Above: **A lunar photograph taken during December 1972 by Apollo 17, NASA's last lunar manned flight to date.** ***Below:*** **Crater Kepler and vicinity, on the moon's northern limb, as photographed by Astronaut Richard Gordon during Apollo 12 in November of 1969.** ***At right:*** **An earth portrait taken by Apollo 13 in April of 1970.**

due to a greenhouse effect in which Venus' mostly carbon-dioxide atmosphere admits sunlight but traps outgoing heat radiation.

Mars, the planet of fabled canals, was found to have no canals but rather an extensive alluvial system from a period much earlier in its history when water perhaps flowed freely. It has an atmosphere, only about 1/100 the density of earth's, made up mostly of carbon dioxide. Examinations of its soil reveal no signs of life. The planet's water is locked in its north polar ice cap and in a subsurface tundra. The tenuous atmosphere would turn any escaping liquid water promptly into vapor. Mars' surface is drier than the driest deserts on earth.

NASA spacecraft studies of Mars and Venus suggest that once they had rivers and seas and that their atmospheres and temperatures were benign. They cannot determine whether life developed during these benign periods. Spacecraft could detect no intrinsic magnetic fields around Mars or Venus. Surprisingly, a faint magnetic field was discovered around Mercury. This discovery raised questions about theories of magnetic field origin which call for rapid rotation and liquid metal interior. Mercury rotates slowly. Its surface is pocked with ancient craters, much like the moon's, and it is airless except for wisps of noble gases.

Close-range NASA spacecraft observations of the giant planets Jupiter and Saturn revealed they have no solid surfaces. Beneath their deep turbulent clouds, the planets are composed mostly of liquid hydrogen. They have vast magnetic fields that apparently emanate from their rapid rotations and metallic liquid hydrogen in their interiors. The temperatures and pressures that are responsible for metallic liquid hydrogen are impossible to duplicate on earth. Like earth's, their magnetic fields are blown by the solar

Mariner 10 returned 8000 photographic images of the planet Mercury in its two flybys of 1974. *Above right* is a Mariner 10 view of Mercury's northern limb. *Below:* Sunrise on Venus, as seen by a Pioneer Venus Orbiter's cloud polarimeter. *At far right:* A Pioneer-Venus cloud cover photograph, showing Venus' wind patterns.

Below: Mars, photographed by a Viking spacecraft. *Above:* This is a picture of the planet's surface, taken by the Viking 2 Lander on its 1050th day on Mars.
At above far right is a Viking 2 Lander photo of another sector of its Utopia Planitia landing site, showing the water-ice 'snow' that typifies winter on Mars.

wind into vast teardrop shapes. Also like earth's, their magnetic fields have trapped atomic particles creating regions of concentrated radiation far larger and more intense than earth's. NASA spacecraft data led to discovery of many small satellites around Jupiter and Saturn and a ring around Jupiter. With this ring discovery and the discovery by a NASA airborne observatory of rings around Uranus, only Neptune of the giant outer planets is not known to be ringed.

The satellites of Jupiter and Saturn appear as points of light in earth telescopes, but NASA's spacecraft sweeping nearby the planets telecast close-ups not only of the planets but also of their satellites. With these telecasts, for the first time, people saw the surfaces of many of these satellites.

Jupiter has four large Galilean satellites, named for their discoverer, the 17th-century Italian astronomer Galileo, and a dozen smaller ones. One of the four, Ganymede, is the largest satellite in the solar system and is larger than the planets Mercury or Pluto. Its surface shows signs of progessive change before it froze solid more than 3 billion years ago. Another, Callisto, shows scars made by ancient meteorite bombardments. Its surface appears to have changed little in four billion years. Callisto, like Ganymede, has a surface mix of rock and water ice and is larger than the planets Mercury or Pluto. Europa, about the size of earth's moon, has an ice-covered surface that is the smoothest seen on any celestial body. All three satellites are made up of substantial quantities of water ice.

Until it was viewed close-up, astronomers had believed that Io, the remaining Galilean satellite, was dry, dead and rocky like our moon. Spacecraft revealed that Io was rocky and waterless and had active volcanoes.

Below: **Jupiter's largest moon, Ganymede, as photographed by Voyager 1 on 2 March 1979.** ***Below, at bottom:*** **A 4 March 1979 Voyager 1 photo of the Jovian moon Io.** ***At right:*** **A Voyager view of the giant planet Jupiter, with a 'star background' overlay.** ***Overleaf:*** **A Voyager 1 image of Saturn from the craft's 1980 flyby.**

NASA spacecraft confirmed that water ice makes up most of Saturn's rings and all or a large part of Saturn's satellites. The major exception is the outermost satellite, Phoebe, which close-up pictures suggest is a captured outer solar system asteroid.

Earth telescopes show vast gaps in Saturn's rings. Spacecraft close-ups reveal that the gaps are filled with many additional very thin ringlets.

Saturn's satellite Titan, larger than Mercury or Pluto, was once considered the largest satellite in the solar system. The Voyager 1 spacecraft close-up measurements showed it was slightly smaller than Gamymede.

However, Titan is of interest because it is the only satellite in the solar system with a substantial atmosphere. Spacecraft studies indicate that the atmosphere is more than eight percent nitrogen with methane, ethane, acetylene, ethylene, hydrogen, cyanide and other organic compounds making up the rest. Its atmospheric surface pressure is about 1.6 earth's sea-level pressure. Its atmosphere resembles that which earth is presumed to have had in primeval times. A smog apparently caused by chemical reaction of sunlight on methane in Titan's atmosphere envelops the satellite.

Voyager 1 further determined that Titan is composed of about equal amounts of rock and water. Scientists observed that with suitable temperature Titan's environment could trigger pre-life chemistry. Some speculated that the smog-generated greenhouse effect could have accumulated enough heat over the last few billion years to raise the temperature to satisfactory levels. However, Voyager 2 took Titan's surface temperature and found it was a chilling 95° Kelvin (-288° Fahrenheit), far too cold for water to liquefy or for significant progress in pre-life chemistry. Additional information is presented in the discussions that follow on each type of spacecraft. Together, they provide but a glimpse of the infinite opportunities for the advancement of knowledge offered by the exploration of space.

At right: **This is a computer color-enhanced Voyager 2 photo of a part of Saturn's rings. Saturn's satellite Titan is the only satellite in the solar system known to have a substantial atmosphere. *Below* is a true color Voyager photograph of Titan, and *at bottom, below* is a false color photo of Titan's super-atmospheric 'haze.'**

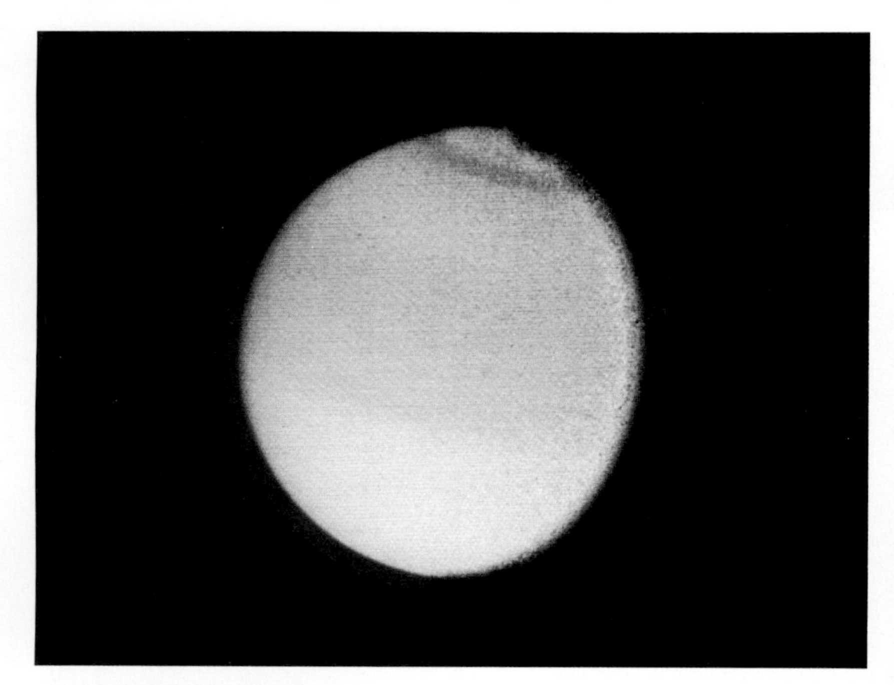

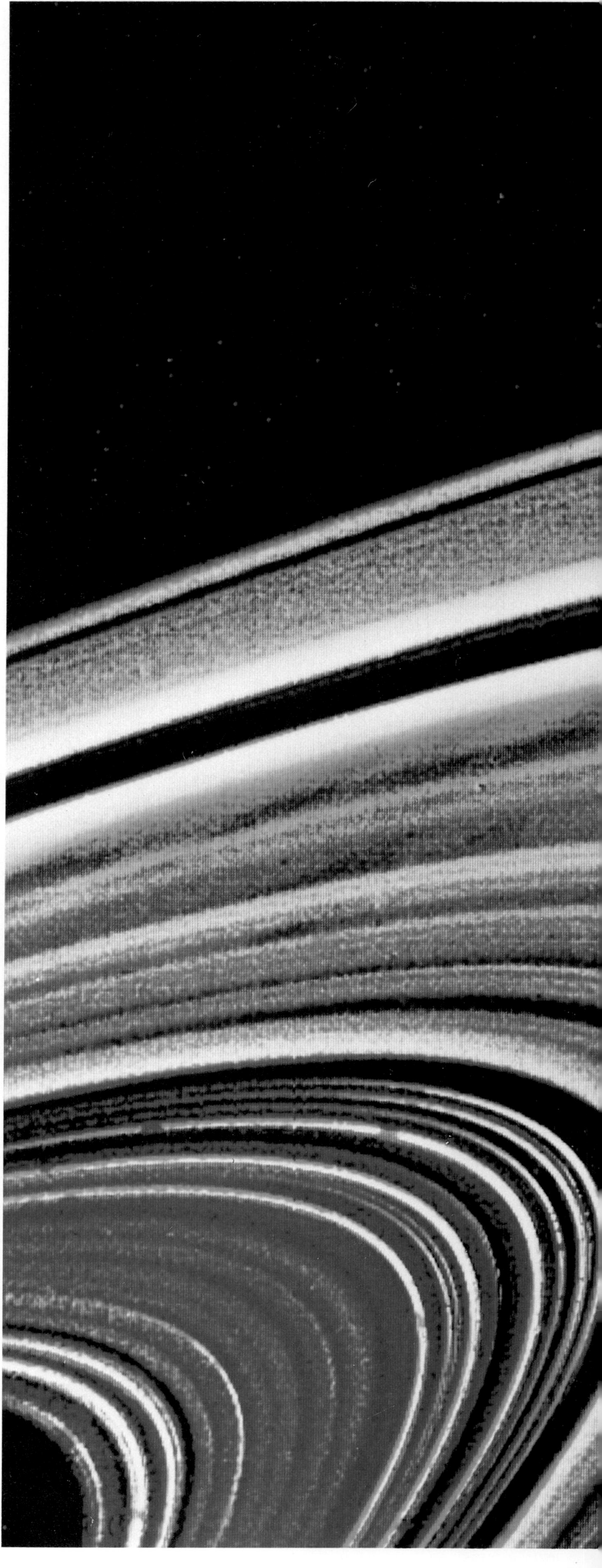

Explorers

Since 1 February 1958, when two organizations that later became part of NASA launched Explorer 1, NASA has launched more than 60 Explorers. A series of satellites that is comparatively small and vary in size and shape, it carries a limited number of experiments into the most suitable orbits.

The Explorer designation has been assigned not only to small satellites conducting space science missions but also to those whose primary mission is in either satellite applications or technological research. Explorers devoted principally to space science programs are described in this section.

NASA's launch of Explorer 11 on 27 April 1961 was the first step of its long range program to probe the universe's secrets that are veiled by earth's atmosphere. Designed to monitor gamma rays, the satellite's data appeared to contradict the steady state theory of constant destruction and creation of matter.

NASA and the US Navy launched Explorers 30 and 37 on 19 November 1965, and 5 March 1965, respectively. The satellites monitored solar X- and ultraviolet rays during periods of declining and increasing solar activity.

Perhaps the longest satellites ever launched were NASA's Radio Astronomy Explorers 38 and 49. The tiny bodies of these satellites were each crossed with two antenmas that were about three times as long as the Washington Monument is high. The antennas were unreeled to form a vast 'X' in space where they received natural radio signals that do not ordinarily reach earth, thus filling a gap in our radio astronomy knowledge.

Explorer 38, launched on American Independence Day 1958, surprised astronomers by reporting that earth sporadically emits natural radio waves. Until then, the only planet known to emit radio waves was Jupiter. Earth was so noisy that it drowned out many other sources. However, Explorer 35 was also able to report that the sun emitted more low frequency radio signals than scientists anticipated.

Earth's radio noise prompted NASA to make Explorer 49, the second Radio Astronomy Explorer, into a lunar-anchored satellite. After the 10 June 1973 launch, Explorer 49 was maneuvered into lunar orbit, far enough away to prevent radio interference from earth.

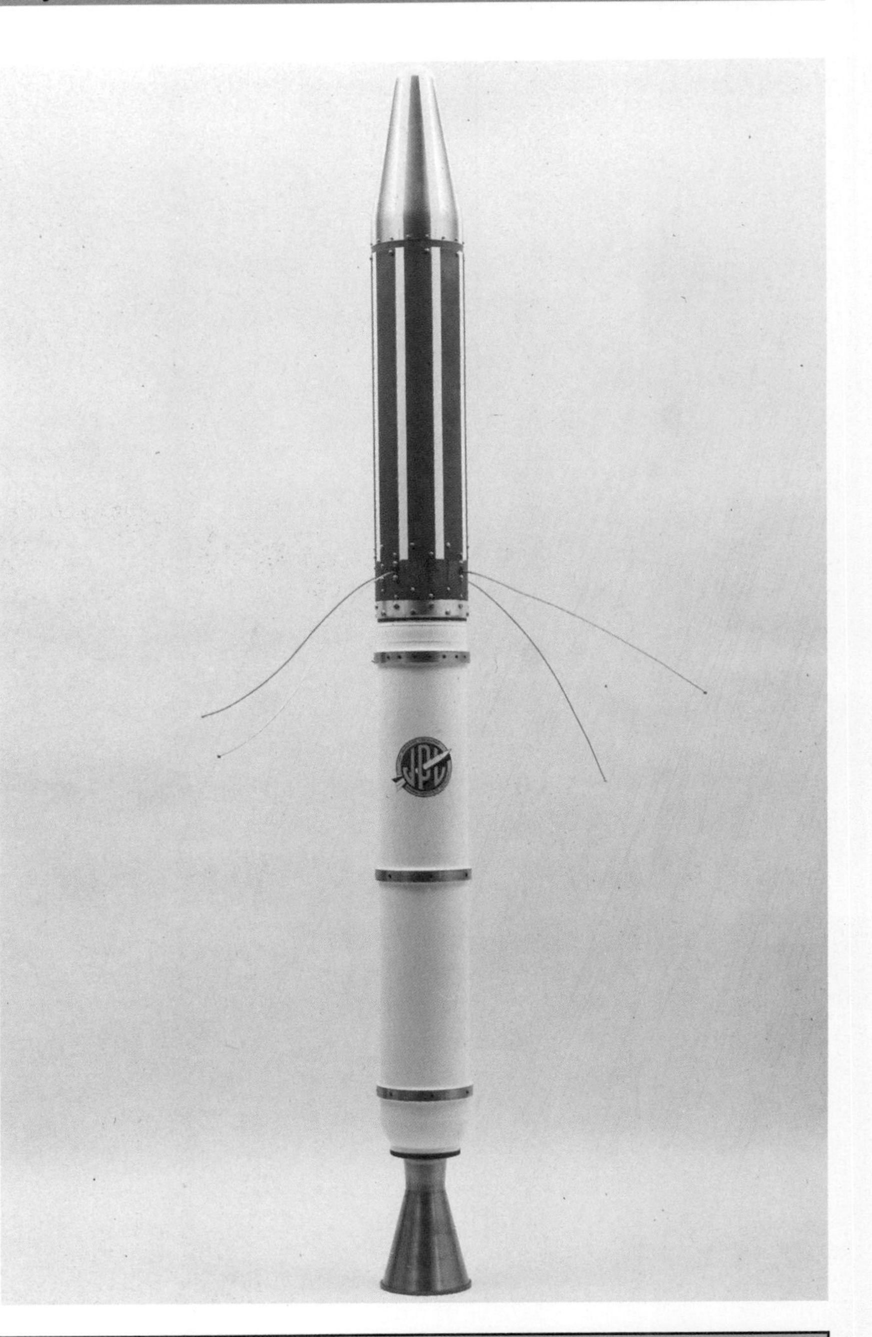

At right and below: Explorer 1, America's first satellite. Launched on 1 February 1958, it transmitted data until 23 May, and was to remain in orbit for 12 years. Among Explorer 1's accomplishments was the confirmation of the existence of the Van Allen Radiation Belt, located at 600 miles altitude. _At far right:_ A Jupiter C launch vehicle—a successful change from the Vanguard—carries Explorer 1 aloft.

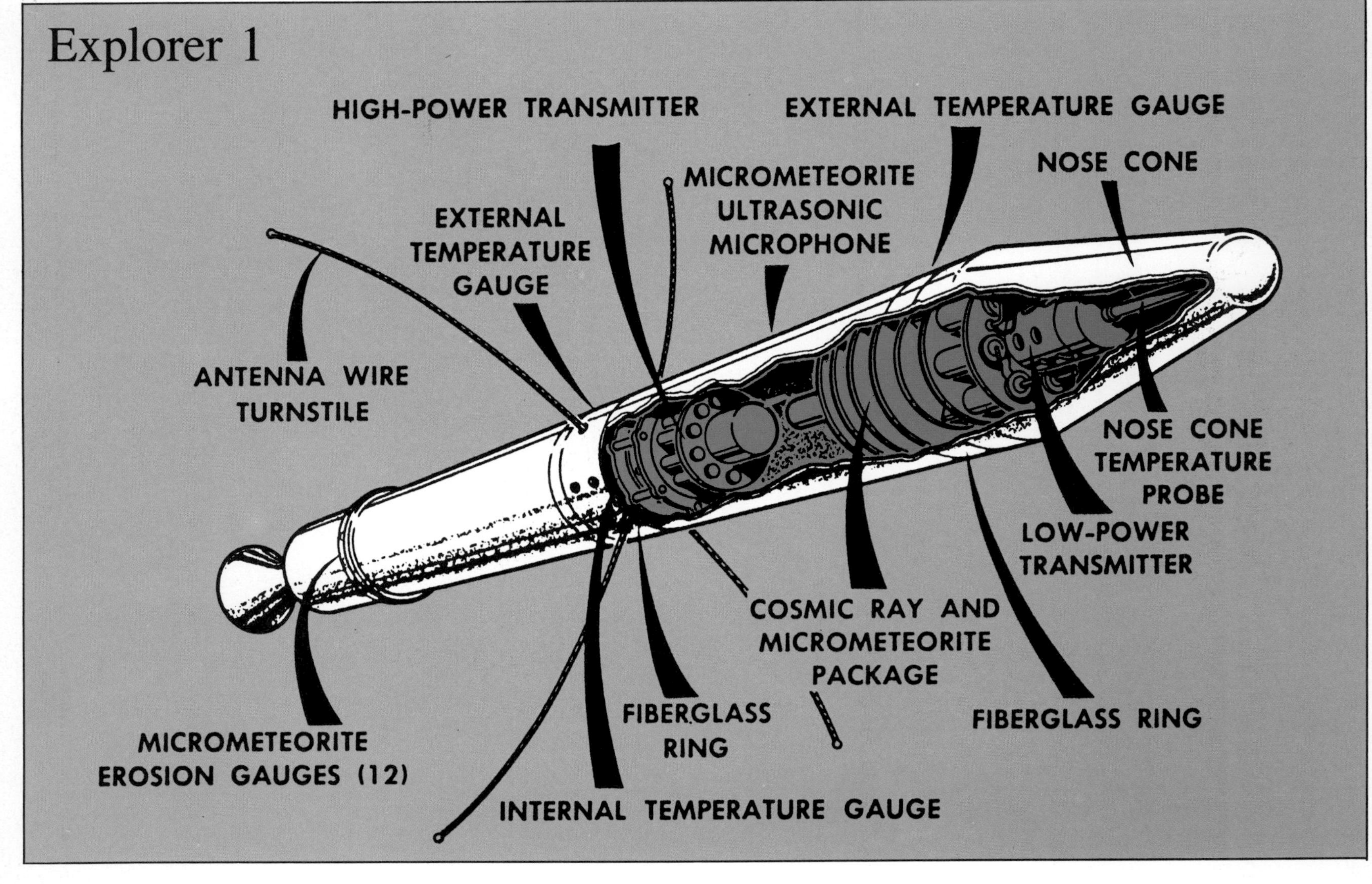

Explorer 42, launched 12 December 1970, was the first of a new category of NASA spacecraft called Small Astronomy Satellites (SAS). Designed to pick up X-rays, it gathered more data in a day than sounding rockets accumulated in the nine previous years of X-ray astronomy. Astronomers used its data to prepare a comprehensive X-ray sky map and X-ray catalog. Data from Explorer 42 suggested that superclusters of galaxies may be bound together by tenuous gases whose total mass is greater than that of the optically visible galaxies. This would provide a significant percentage of the mass needed to support the theory that our expanding universe will eventually contract.

Explorer 42 was the first NASA satellite launched by a foreign nation. Italy launched Explorer 42 from its floating San Marco platform in the Indian Ocean off the coast of Africa, near Kenya. Because Explorer 42 was launched on Kenya's independence day, it was also named *Uhuru* which is Swahili for 'freedom.' On 15 November 1972, Italy launched NASA's Explorer 48, the second SAS, from its San Marco platform. Explorer 48 continued the expansion of knowledge about gamma ray sources first begun by Explorer 11. It provided data that could be interpreted as supporting the theory that the universe is composed of regions of matter and antimatter.

Explorer 53, the third SAS, was launched from San Marco on 7 May 1975. It discovered many additional X-ray sources, including one identified as a quasistellar object (quasar) only 783 million light years away. This is the closest quasar yet discovered.

Many new discoveries were made by the International Ultraviolet Explorer (IUE), launched 26 January 1978. IUE is a joint project of NASA, the United Kingdom and the European Space Agency. IUE data supported a theory that a black hole with the mass of a thousand solar systems existed at the center of our Milky Way galaxy and revealed that our galaxy had a halo of hot gases. The data provided evidence that so-called twin quasars were actually a double image of the same object. Light waves from the quasar are bent around a massive elliptical galaxy which acts as a gravitational lens to produce the double image picked up by ground observatories.

Atmosphere Explorers have confirmed or redrawn our conceptions of earth's tenuous upper atmosphere. The first, Explorer 8, was launched 3 November 1960. It confirmed that temperatures of electrons in the upper ionosphere are higher by day than by night. It discovered that oxygen predominates in the ionosphere up to an altitude of about 650 miles where helium predominates. A secondary experiment indicated that micrometeoroid quantities varied inversely with size.

Air Density Explorers 9, 24 and 39 were launched 16 February 1961; 21 November 1964; and 8 August 1958, respectively. These were essentially 12-foot balloons of aluminum foil and plastic laminate that were inflated in orbit. Air drag on the satellite indicated air density. The satellites revealed that atmospheric density varied from day to night, with the 27-day rotation period of the sun, with the 11-year solar cycle, and with violent eruptions on the sun.

Explorer 25 was launched on the same Scout booster that orbited Explorer 24, the first multiple launch by a single vehicle. Explorer 40 was launched with the same Scout vehicle as Explorer 39. They demonstrated a correlation between air density and solar radiation. Explorers 25 and 40 were called Injuns and University Explorers because they were built by the University of Iowa.

Explorers 17 and 32, launched 2 April 1963, and 25 May 1956, gathered information about the composition of neutral atoms and molecules. Explorer 17 confirmed Explorer 8 indications of a belt of neutral helium in the upper atmosphere. In radio-echo soundings of the ionosphere, radio signals at different frequencies were transmitted from the ground. The reflected frequency discloses electron density; the return time indicated the altitude or distance at which the density was encountered. Ground-based sounding cannot provide information about the upper ionosphere because electron density increases up to a certain altitude and then tapers off. In addition, many areas of earth are too remote or inaccessible for ground-based radio sounding. These problems were solved by using satellites as topside sounders. They beamed radio waves into the ionosphere from altitudes far above the region of maximum electron density.

NASA topside sounders were Explorer 20, launched 25 August 1964, and Explorer 31, launched 28 November 1965. Scientists correlated data from these satellites with data from the Canadian topside sounders Alouette 1 and 2. Explorer 22, an ionosphere beacon launched 10 October 1964, also

At right: **A four-stage Scout vehicle, laden with the US Air Force Instrumented Test Vehicle, at Launch Pad 3 on Wallops Island.** ***At far right:*** **The Explorer 38 Radio Astronomy Telescope package is shown being mated to the third stage of its Delta launch vehicle.**

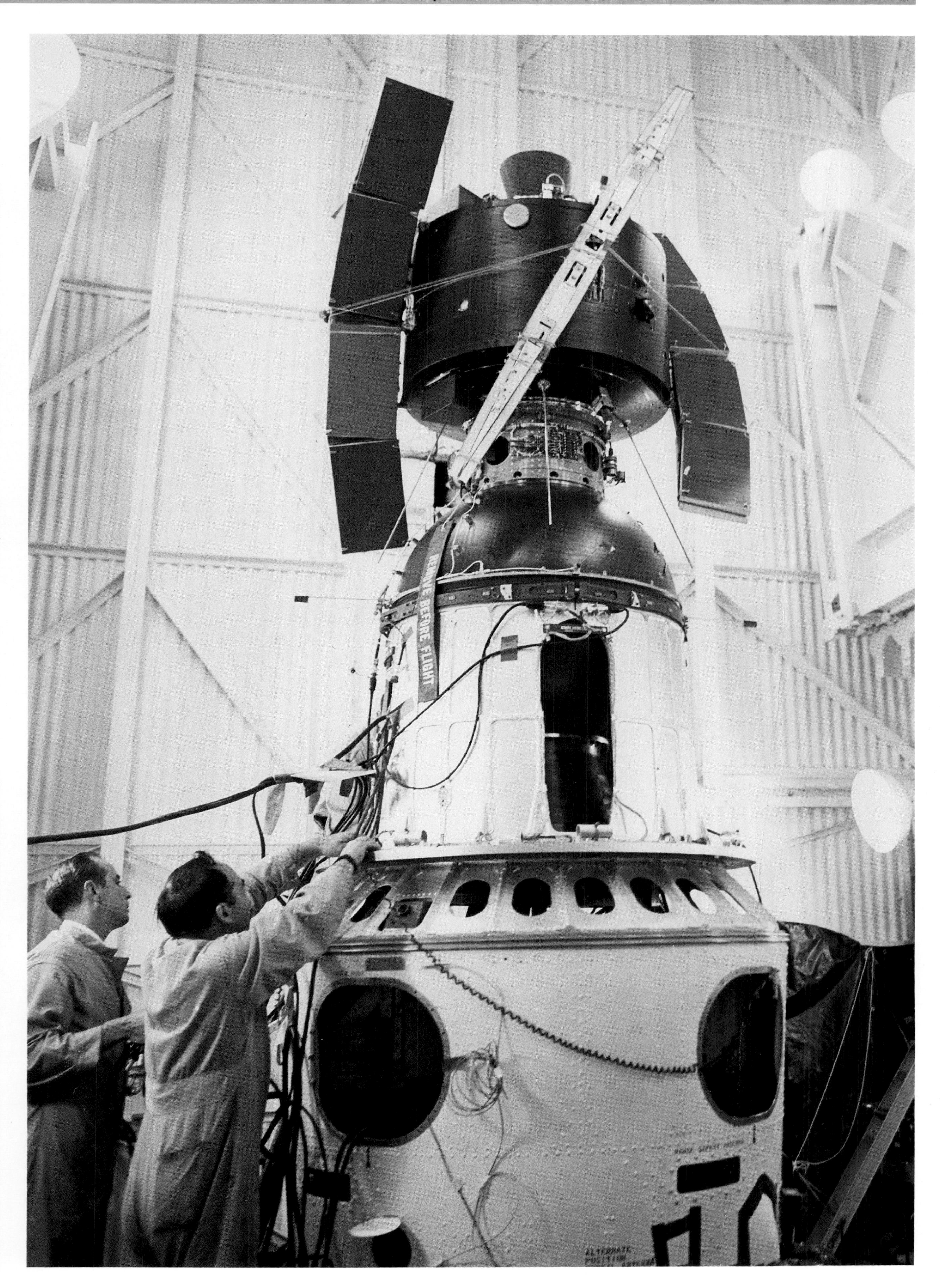
BEFORE FLIGHT
ALTERNATE
POSITION

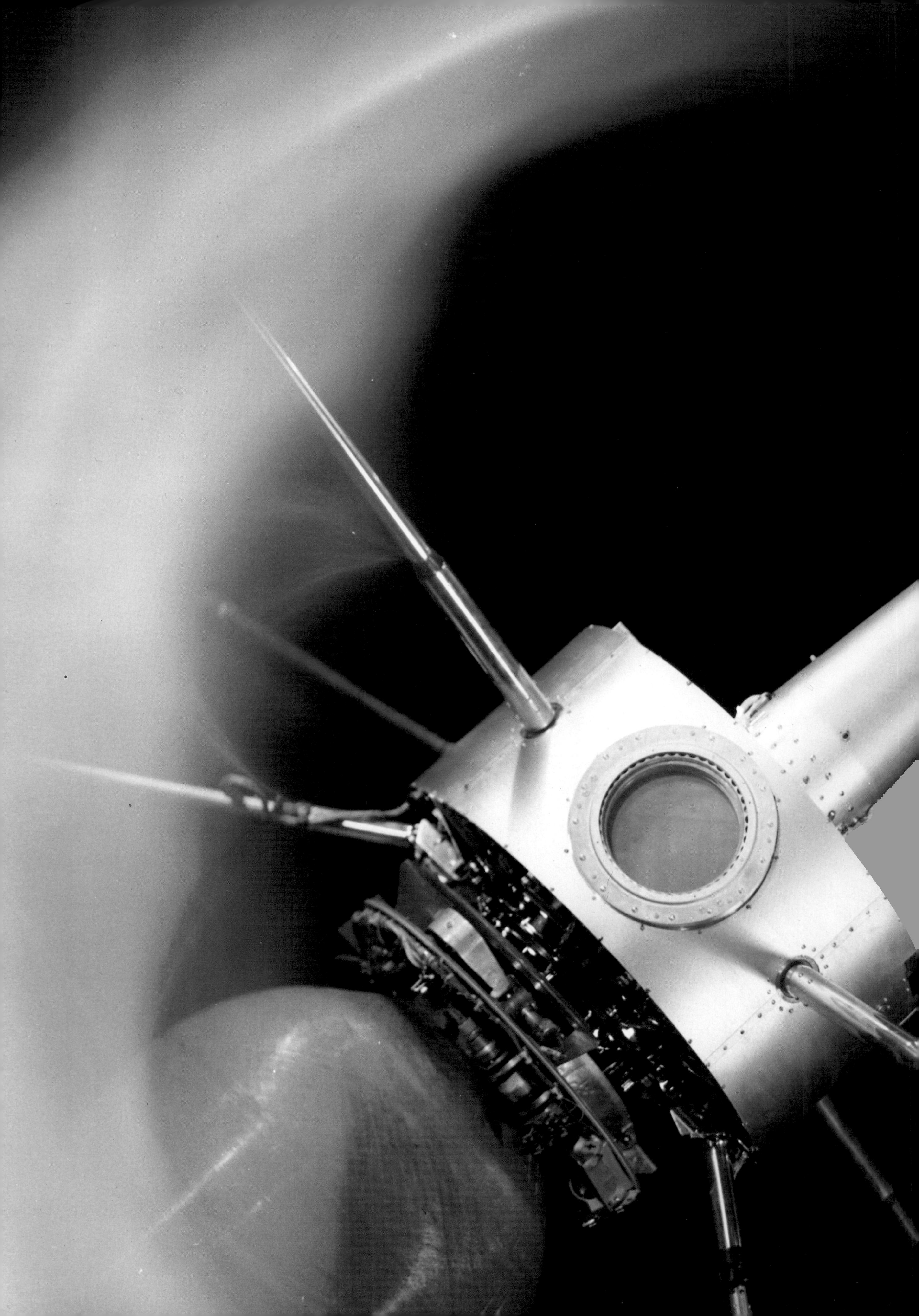

These pages: Swathed in some imaginative artist's version of 'the aethers,' Explorer 20 is shown with antenna arms outspread. The sphere atop the craft's truncated cone is an experiment sphere. This craft was launched from Vandenberg Air Force Base, with Scout launch vehicle, on 25 August 1964. Explorer 20 made radio soundings of the upper atmosphere, and collected data on the earth's ionosphere.

measured electron density in the ionosphere. Explorer 22 was built with quartz reflectors for the first major experiment in laser tracking.

The thermosphere, a region of the upper atmosphere, was believed to be relatively stable until Explorers 51, 54 and 55 were launched. These satellites were equipped with onboard propulsion systems that enabled them to dip deep into the atmosphere and pull out again, taking measurements and providing extensive data about the upper thermosphere. They found that the thermosphere behaved unpredictably with winds 10 times stronger than normally found at earth's surface. They discovered abrupt and constantly changing wind shears. Their data contributed significantly to knowledge about energy transfer mechanisms and photochemical processes (such as those that create the ozone layer) in the atmosphere. Launch dates for the three Explorers were 16 December 1973; 6 October 1975; and 19 November 1975.

The Solar Mesosphere Explorer, launched 6 October 1981, provided comprehensive data on how solar radiation creates and destroys ozone in the mesosphere, an atmospheric layer below the thermosphere and above the stratosphere. The University of Colorado designed and built the Solar Mesosphere Explorer and operated it for a year after launch.

Geophysical Explorers

NASA's first successful satellite launch was Explorer 6, orbited 7 August 1959. Explorer 6 added to information about the Van Allen Region and micrometeoroids. It also telecast a crude image of the north Pacific Ocean.

Shown *below* in vibration testing, is Explorer 48 (aka Small Astronomy Satellite B), with solar panels folded in launch configuration. *At right:* A Scout launch vehicle rises aloft from its launch pad at Vandenberg Air Force Base. Explorer 51 (Atmosphere Explorer C), which investigated earth's thermosphere, is shown at *far right.*

Explorer 7, launched 13 October 1959, provided data revealing that the Van Allen Region fluctuated in volume intensity and suggested a relationship of the region with solar activity. It indicated that variations in solar activity may also be related to the abundance of cosmic radiation in earth's vicinity, magnetic storms and ionospheric disturbances.

Explorer 10, launched on 25 March 1961 to gather magnetic field data, was the first spacecraft to obtain information that suggested that the interplanetary magnetic field may actually be an extension of the sun's field carried outward by the solar wind.

With Explorer 12, launched 15 August 1961, scientists were able to arrive at many conclusions about space: the Van Allen Region is a single system of charged particles rather than several belts; earth's magnetic field has a distinct boundary; the solar wind compresses the earth's magnetic field on the sun's side and blows it out on the other; and geomagnetic storms that cause radio blackouts and power outages may result from solar flares.

On 2 October 1952, NASA launched Explorer 14 to monitor the Van Allen Radiation Region during a period of declining solar activity. Scientists in the meantime discovered that the United States project Starfish, involving a high altitude nuclear burst in July 1962, had created another artificial radiation belt. Explorer 14 was joined by Explorer 15 on 27 October to help monitor this belt. The two satellites' data helped ease scientific anxiety by indicating that atomic particles making up the belt were rapidly decaying.

NASA launched Explorer 26 on 21 December 1964, and its data increased understanding of how atomic particles traveling toward earth from outer space are trapped by earth's magnetic field and how they spiral inward, along earth magnetic field lines in the northern and southern latitudes, interacting with the atmosphere to generate auroras.

Explorer 45, launched 15 November 1971, further investigated the relationships of geomagnetic storms, particle radiation and auroras. It was the second satellite launched by an Italian crew from the San Marco platform off Kenya in the Indian Ocean.

NASA's Interplanetary Monitoring Platforms, or IMP Explorers, added significantly to knowledge about how earth's magnetic field and the Van Allen Radiation Region fluctuate during the 11-year cycle. IMP Explorer 18 confirmed that earth's magnetic field was shaped like a giant cosmic teardrop. It discovered a shockwave ahead of the earth's field. The shockwave is caused by impact of the speeding solar wind with the earth's

Below: **Another view of Explorer 51. This craft carried out 15 experiments, including studies of photochemical/ultraviolet reactions. The 35-inch sphere shown *at right* is Explorer 17 (Atmospheric Structure Satellite S-6), which measured the density, composition, pressure and temperature of earth's atmosphere.**

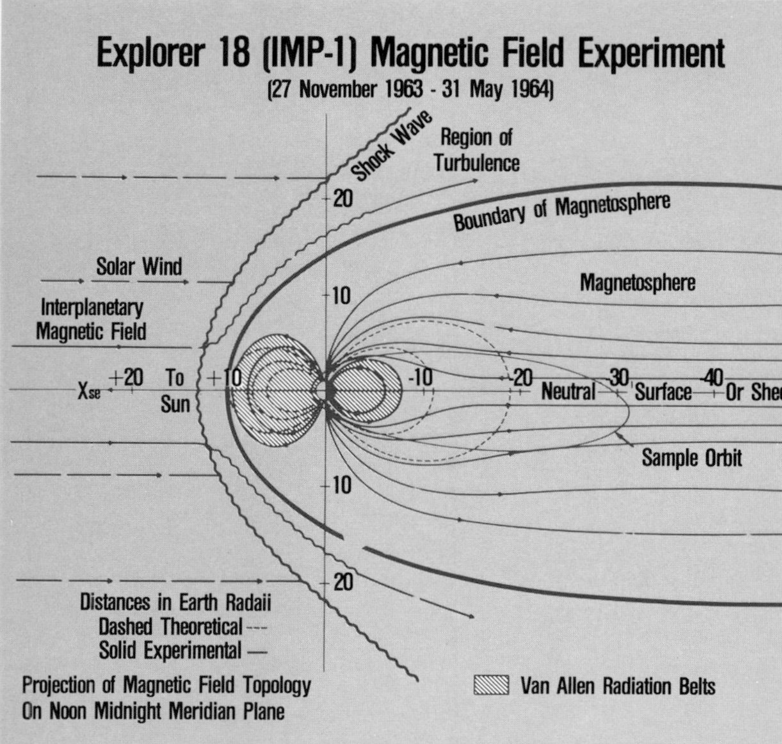

Shown at right is Interplanetray Monitoring Platform satellite Explorer 18 (IMP-1). Explorer 18 measured cosmic radiation, magnetic fields and the solar wind in interplanetary space, and discovered a high energy radiation region beyond the Van Allen Belt. *Above* is a diagram of the Explorer 18 orbit and environment.

field. Between the shockwave and the magnetopause, or magnetic field boundary, Explorer 18 discovered a turbulent region of magnetic fields and atomic particles.

IMP Explorer 33 was the first satellite to provide evidence that the geomagnetic field on earth's night side extends beyond the moon. IMP Explorer 35, a lunar-anchored (lunar orbiting) IMP, gathered data about micrometeoroids, magnetic fields, the solar wind and radiation at lunar altitudes. Its instruments revealed the moon to be what one scientist termed a 'cold nonmagnetic nonconducting sphere.'

A two-satellite Dynamic Explorer project, launched simultaneously on 3 August 1981, significantly contributed to data on coupling of energy, electric currents, electric fields, and plasmas (hot electrified gases) between earth's magnetic field, the ionosphere, and the rest of the atmosphere. Among their discoveries were nitrogen ions in the geomagnetosphere. They also confirmed existence of the polar wind which is an upward flow of ions from the polar ionosphere.

The Dynamic Explorers complemented studies of the three International Sun-Earth Explorers (ISEE), a joint project of NASA and the European Space Agency. ISEE 1 and 2 were launched into earth orbit on 22 October 1977. ISEE 3 was placed in a heliocentric orbit near the sun-earth libration point.

The ISEE program focused on solar-terrestrial relationships as a contribution to the International Magnetospheric Study. The three spacecraft obtained a treasure trove of new information on the dynamics of the geomagnetosphere, the transfer of energy from the solar wind and energization of plasma in the geomagnetotail. For example, ISEE 1 found ions from our ionosphere accelerated in the goemagnetotail to fairly high energies. Previously, scientists thought these high energy particles originated from the solar wind.

In 1982, when its mission was completed with fuel to spare and all imstruments in working condition, ISEE 3 was put through a series of complex maneuvers to explore the earth's magnetotail through December 1983, to fly across and study the wake of Comet Giacobini-Zinner in September 1985, and to observe from comparatively close range the effects of the solar wind on Halley's Comet in late 1985 and early in 1986.

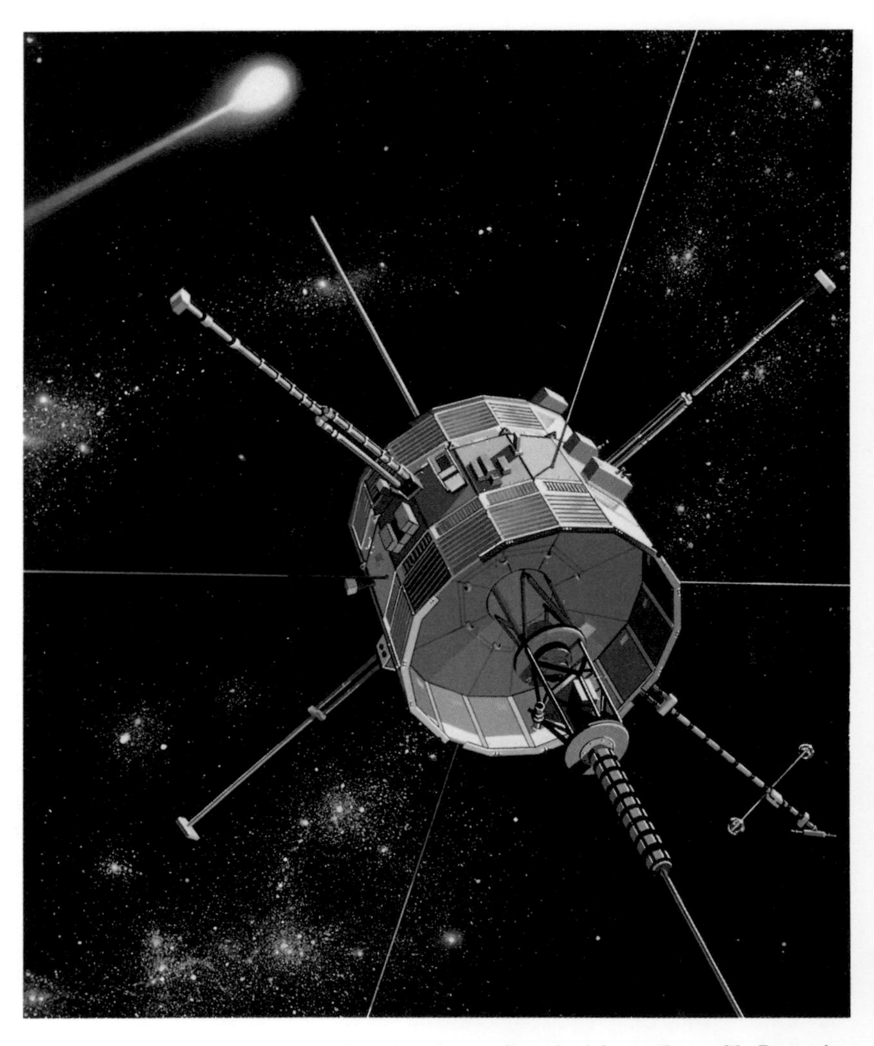

Below: **Dynamics Explorer-A (DE-A) (near) was launched in tandem with Dynamics Explorer-B (DE-B) (far)—see text, this page.** ***Above:*** **An artist's conception of the ISEE-3 spacecraft en route to its encounter with comet Giacobini-Zinner.** ***At right*** **is Explorer 50, the last of the IMP spacecraft, launched on 25 October 1973.**

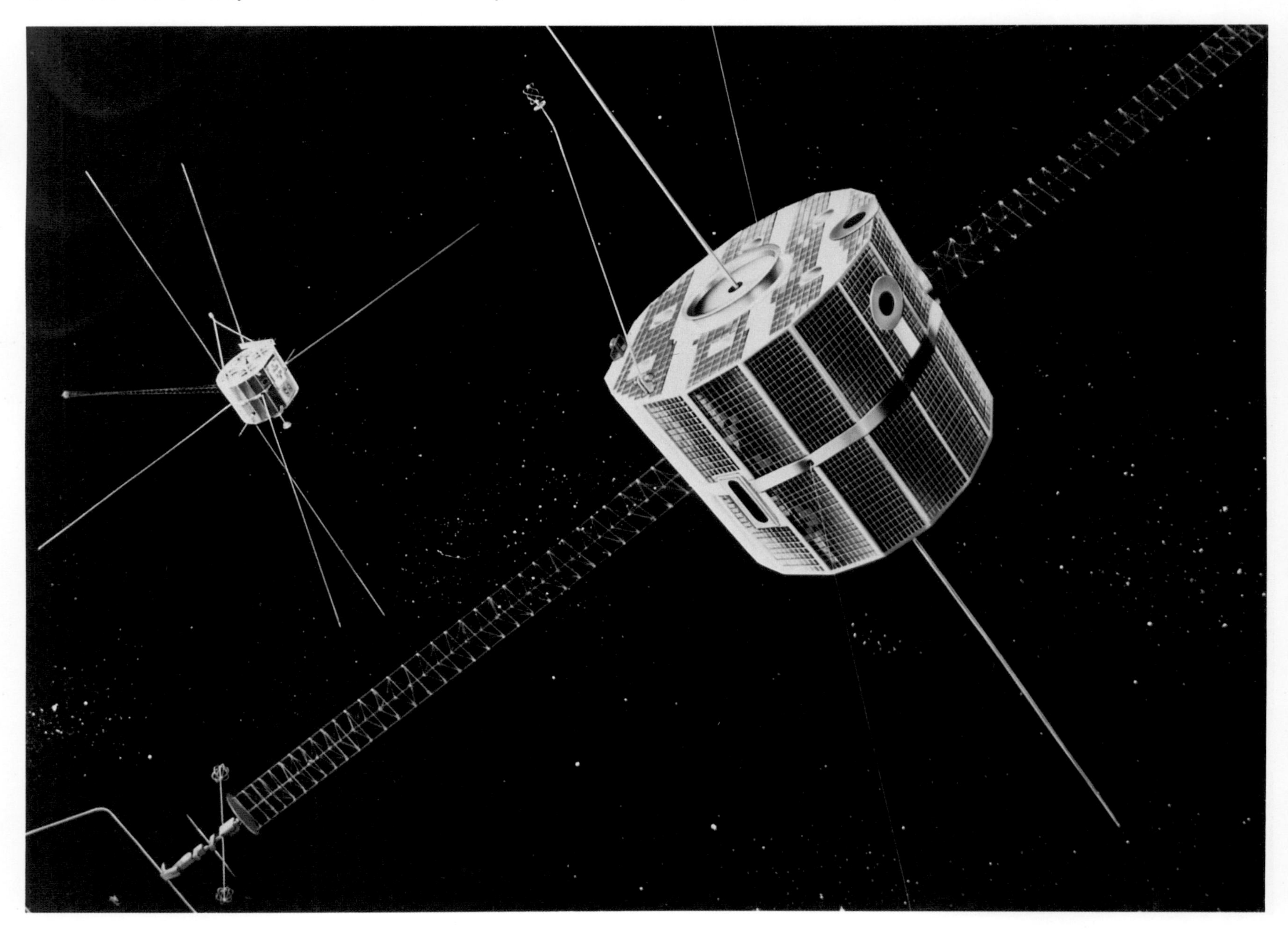

Astronomical Observatories

Man's view of the universe is narrowly circumscribed by the atmosphere which blocks or distorts most kinds of electromagnetic radiations from space. Analyses of these radiations (radio, infrared, visible light, ultraviolet, X-rays and gamma rays) give important new information about the phenomena in our universe.

The small Astronomical Explorers indicated the great potential for acquiring new knowledge by placing instruments above earth's obscuring atmosphere. Consequently, NASA orbited a series of large astronomical observatories bearing a great variety of instruments which have significantly widened our window on the universe.

The first successful large scale observatory was Orbiting Astronomical Observatory 2, nicknamed Stargazer, which was launched 7 December 1968. In its first 30 days, OAO 2 collected more than 20 times the celestial ultraviolet data acquired in the previous 15 years of sounding rocket launches.

Among the volumes of data provided by OAO 2 are the following discoveries:

- Stars that are many times as massive as our sun are hotter and consume their hydrogen fuel faster than estimated on the basis of ground observations. OAO 2 data contributed to resolving a disparity between observations made from the ground and theories of stellar evolution.
- Another stellar theory was brought into question. According to this theory, intensity of celestial objects should be less in ultraviolet light than in visible light. However, several galaxies that looked dim in visible observations from earth were bright in ultraviolet observations by Stargazer.
- Diffuse dust nebulae are regarded by many astronomers as the location for the formation of stars.
- Stargazer was able to observe Nova Serpentis in 1970 for 60 days after its outburst. It confirmed that mass loss by the nova was consistent with theory.
- Stargazer observations of the Comet Tago-Sato-Kosaka supported the theory that hydrogen is a major constituent of comets. It detected a hydrogen cloud as large as the sun around the comet. Because of our atmosphere, this hydrogen cloud could not be detected by ground observatories.
- Looking toward earth, Stargazer reported that the hydrogen in earth's outer atmosphere is thicker and covers a larger volume than previous measurements indicated.

Launched 21 August 1972, OAO 3 was named for the famed Polish astronomer Copernicus. The satellite provided much new data on star temperatures, chemical compositions and other properties. It continued

At right is a pre-flight view of OAO 3 *Copernicus*, NASA's fourth Orbital Astronomical Observatory. This satellite gathered data on a probable black hole known as Cygnus X-1 (for the intense X-radiation that it emits), which is located in the constellation Cygnus *(shown below)*. Also, see the text above and on page 46.

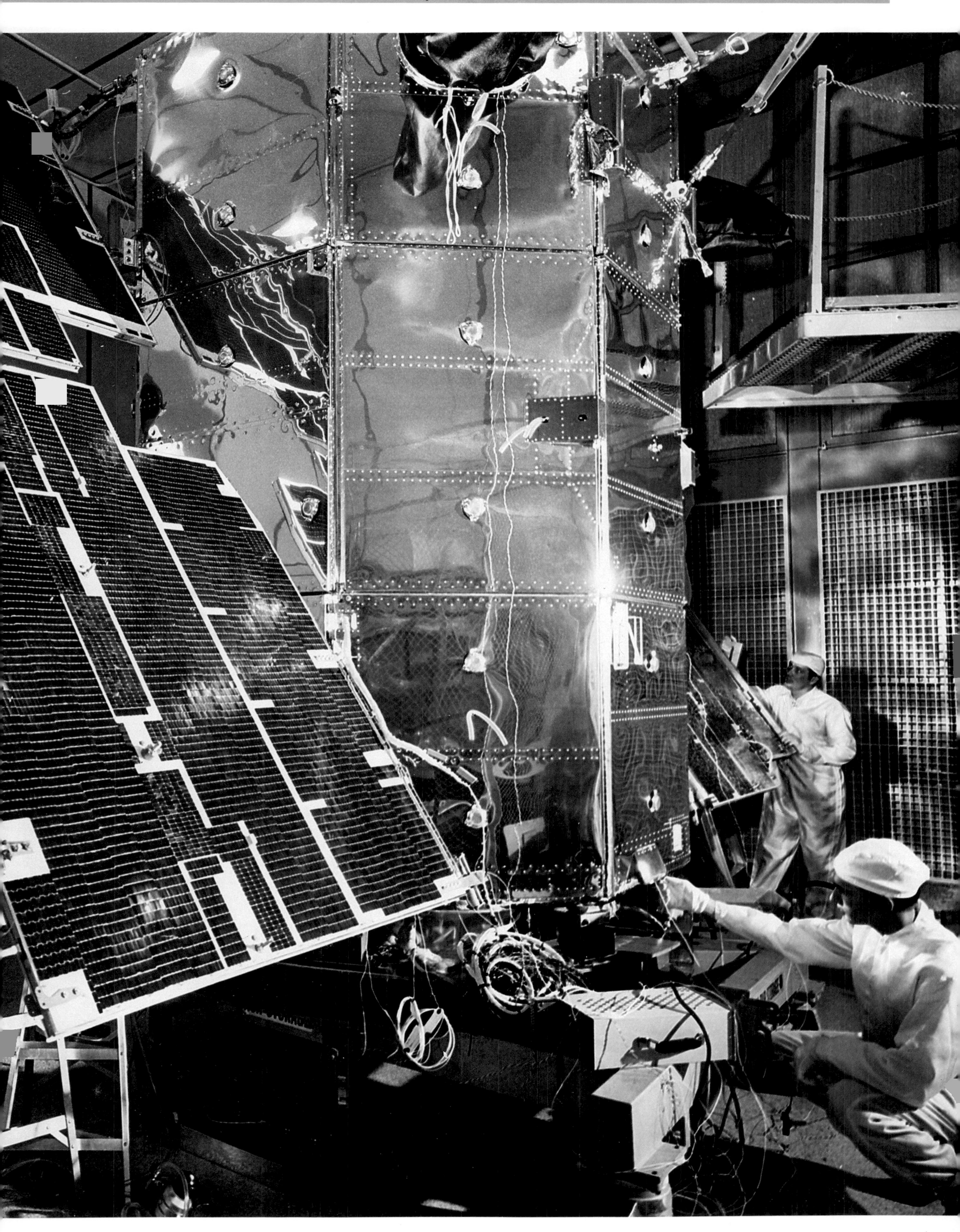

studies of the outer atmospheres of earth, Mars, Jupiter and Saturn. It gathered data on the black hole candidate Cygnus X-1, so named because it is the first X-ray source dlscovered in the constellation Cygnus.

Much of its data supported the hypothesis that Cygnus X-1 is a black hole. A black hole is a one-time massive star that has collapsed to such density that it does not permit even light or other electromagnetic radiations to escape it. Scientists can study Cygnus X-1 because it is part of a binary star system and has a visible companion. In addition, according to theory, a substantial part of the matter dragged into a black hole is transformed into X-rays and gamma rays that are radiated into space before they reach the point of no return.

Among other observations of Copernicus:

- Interstellar dust clouds have fewer heavy elements than our sun. This supported the contention that the sun and planets coalesced from the debris of an ancient supernova.
- Larger amounts of hydrogen molecules than expected were found in interstellar dust clouds.
- Surprisingly large amounts of deuterium (an element with the same atomic number and the same position in the Table of Elements as hydrogen but with twice the mass) were also detected in interstellar dust clouds.

The latter two findings suggest that star formation may be common. The discovery regarding deuterium contradicted a theory that most deuterium, a basic element for atomic fusion in stars, has been exhausted.

High Energy Astronomical Observatory HEAO 2 *Einstein*, launched on 13 November 1978, took this X-ray picture *(below)* of supernova remains in the constellation Cassiopeia. Launched in 1979, HEAO 3 *(at right)* analyzed cosmic ray particles and gamma radiation, discovering a new quasar-like object—see the text on page 48.

High Energy Astronomy Observatories

Three High Energy Astronomy Observatories portray a universe in constant turbulence with components repeatedly torn apart and recombined by violent events.

HEAO 1, launched 12 August 1977, also discovered a new black hole candidate near the constellation Scorpius, bringing the total to four. Other black hole candidates are in or near the constellations Cygnus, Circinus and Hercules. Another major result of HEAO 1 was the discovery of a superhot superbubble of gas 1200 light years in diameter and about 5000 light years from earth. Centered in the constellation Cygnus, the bubble has enough gas to create 10,000 suns. HEAO 1 raised the catalog of X-ray sources from 350 to about 1,500.

The *Einstein* observatory, nicknamed for the famous mathematician, is HEAO 2, launched 13 November 1978. *Einstein* was equipped with more sensitive instruments than HEAO 1. Thus, it was able to discern that the X-ray background observed by HEAO 1 was not coming from diffuse hot plasmas but from quasars. *Einstein* also provided the first pictures of an X-ray burster which is apparently located at the center of a globular cluster called Terzan 2. The bursters are frequently associated with clusters of old stars. They are usually explained in terms of gases interacting violently with neutron stars or black holes, emitting very short bursts of X-rays.

Among other data from *Einstein* are X-ray spectra of supernova remnants. The data support the theory that our system was formed from debris of an ancient supernova.

HEAOs 1 and 2 X-ray measurements related principally to atomic interactions and plasma processes associated with stellar phenomena. HEAO

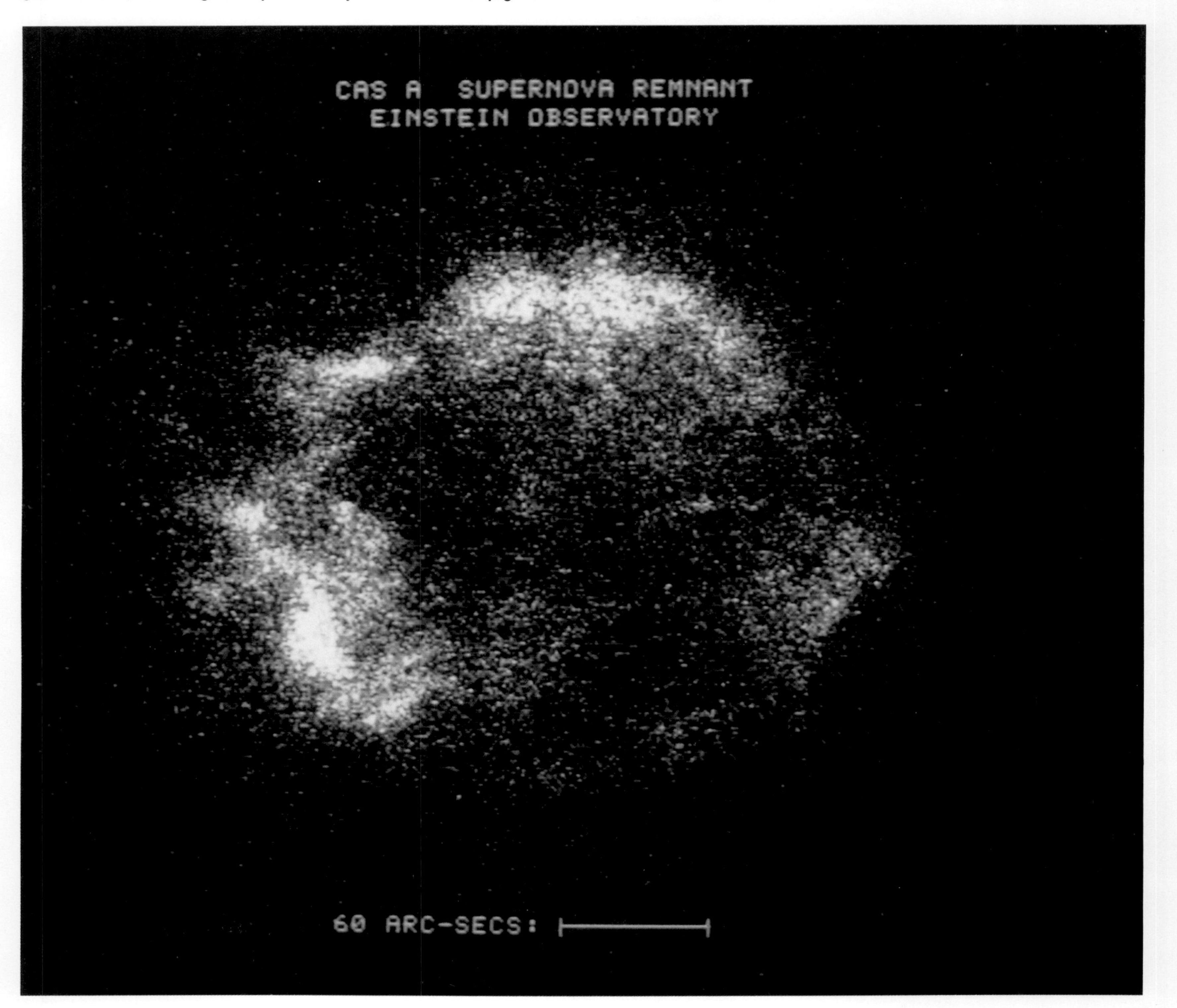

3, launched 20 September 1979, scanned the universe for cosmic ray particles and gamma radiation. The events that HEAO 3 measures result from nuclear reactions in the hearts of stellar objects and the elements they create.

HEAO 3 has observed in the Milky Way's central region gamma rays that apparently emanate from the annihilation of electrons and positrons (the antimatter equivalent of electrons). It has also discovered an object emitting energy in the form of gamma rays equivalent to 50,000 times the sun's total output. The object is 15,000 light years from earth and appears to be undergoing processes on a comparatively small scale that are believed to occur in quasars on a large scale.

The Netherlands Astronomical Satellite (NAS), launched 30 August 1974, was a small X-ray and ultraviolet orbiting observatory. Among discoveries from its data are X-ray bursters, sources that emit bursts of X-rays for seconds at a time. NAS was a cooperative program of NASA and the Netherlands.

Infrared Astronomy Satellite (IRAS)

Launched on 25 January 1983, Infrared Astronomy Satellite revealed many infrared sources in the Large Magellanic Cloud, 155,000 light years from earth, that are not visible from earth, helping scientists compile the first catalog of infrared sky sources.

Because all objects, even cool dark ones that may be the black cinders of dead stars, radiate infrared light, IRAS may discover many other invisible objects. It has revealed stars being born in thick opaque clouds of gas.

IRAS is a joint project of NASA, the Netherlands and United Kingdom. In April 1953, IRAS detected a new comet which came within 3 million miles of earth in May, the closest comet approach in 200 years. It most recently discovered a possible new solar system near the star Vega.

***At near right:* HEAO 2 in preparation. *At below, opposite,* the Infrared Astronomy Satellite (IRAS) is shown being built at the Fokker facility in the Netherlands. *At middle right* is a view of IRAS in the space simulator at NASA's Jet Propulsion Laboratory, and *at far right* is IRAS on its Delta booster at Vandenberg AFB. *Below:* An artist's rendition of IRAS in orbit, after its launch of 25 January 1983.**

6-2
6-L

IRAS, a joint project between NASA, the Netherlands and the United Kingdom. All objects radiate infrared light, a factor that enables the discovery of objects that are too dark or obscured for the eye to see. An example of infrared imaging is *the main photo on these pages*, which is the Andromeda Galaxy M31 as seen through IRAS' infrared apparatus. By comparison, *the inset photo at lower right* is a true color photograph of the same celestial object, taken through the 200-inch Hale Telescope at Mount Palomar Observatory in southern California.

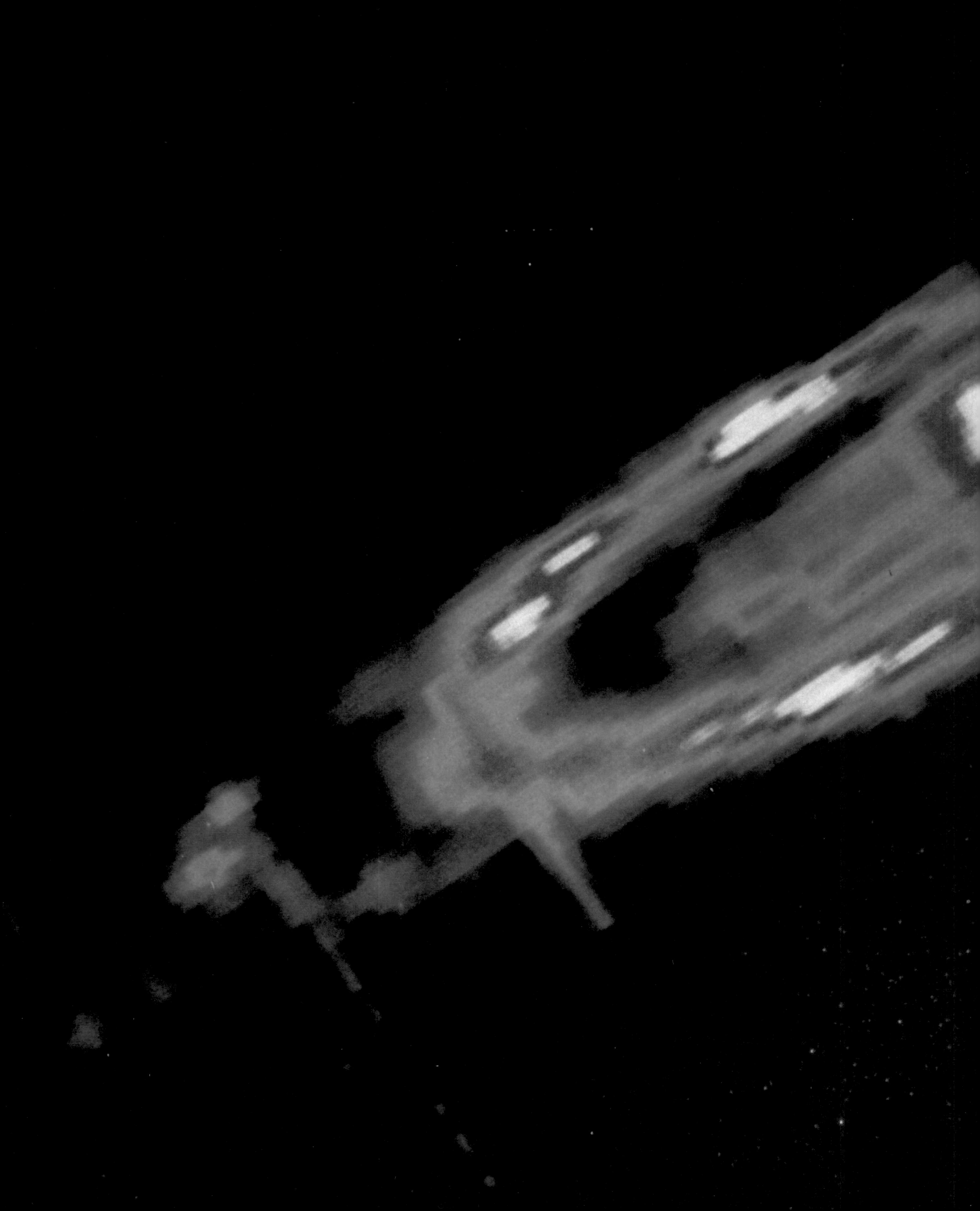

Orbiting Solar Observatories (OSO)

Orbiting Solar Observatories using X-ray and ultraviolet sensors acquired a rich harvest of solar data during the 11-year solar cycle when solar activity went from low to high and then back to low. They photographed for the first time the birth of a solar flare, a great outburst of matter and energy from the sun. When directed toward earth, solar flares can cause black-outs of communications and electricity, magnetic compasses to spin crazily, and enhance displays of the northern and southern lights. OSOs discovered evidence of gamma radiation resulting from solar flares, indicating nuclear reaction in the flares.

OSO 1, launched on 7 March 1962, showed a correlation between fluctuations in temperatures of earth's upper atmosphere and variations in solar ultraviolet ray emissions. OSO 3, launched on 8 March 1967, revealed that the center of our galaxy was the source of intense gamma radiation, which, when confirmed by HEAO 3, led to speculation that matter and antimatter were annihilating each other, leaving energy in the form of gamma rays. The data from OSO 5, launched 22 January 1969, revealed that earth's upper atmosphere may contain as much as 10 times the amount of deuterium (a form of hydrogen with twice its mass) previously estimated.

OSOs discovered and provided information on solar poles where there were cooler and thinner gases than in the rest of the corona. Later, they discovered comparable phenomena on other areas of the sun and scientists named them solar holes. OSO observed and reported on the dramatic coronal transient, a solar explosion hurling out hundreds of thousands of tons of material in huge loops at millions of kilometers per hour.

The Solar Maximum Mission (SMM)

Infinitesimal (.001) reductions in solar energy output that may be related to unusually harsh winters and cool summers on earth were discovered by the Solar Maximum Mission (SMM) satellite. 'Solar Max' was launched 14 February 1980, to study the sun during the high part of the solar cycle. SMM also made the first clear observations of neutrons traveling from the sun to the earth after a flare and confirmed that fusion, the basic process that powers the sun occurs in the solar corona during a flare. A malfunction in January 1981 cut SMM's life short. Solar Max was recovered aboard the Shuttle orbiter *Challenger* during Mission 41-C in April 1984 and returned to Earth for repair.

Orbiting Geophysical Observatories (OGO)

A half dozen Orbiting Geophysical Observatories — launched between September 1961 and June 1965 — have provided more than a million hours of scientific data from about 130 different experiments relating to earth's space environment and sun-earth interrelationships. They significantly

Within the series of nine Orbiting Solar Observatories, the first of two OSO 3 craft failed, to be replaced by the second OSO 3, seen *below.* The last OSO, which was designated OSO 8 and is seen, *at right,* undergoing testing at Hughes Aircraft's El Segundo, California facility. OSO 8 was launched on 21 June 1975.

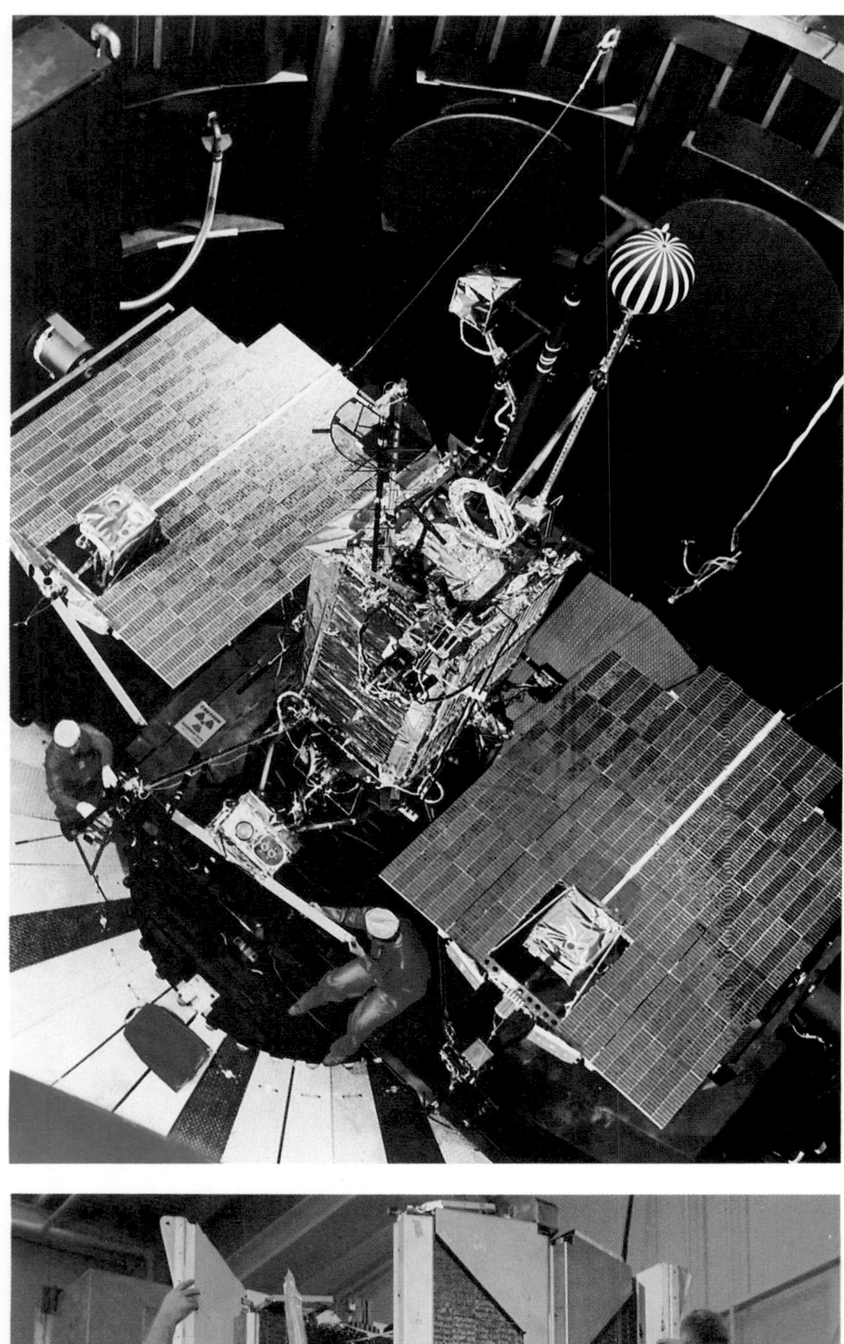

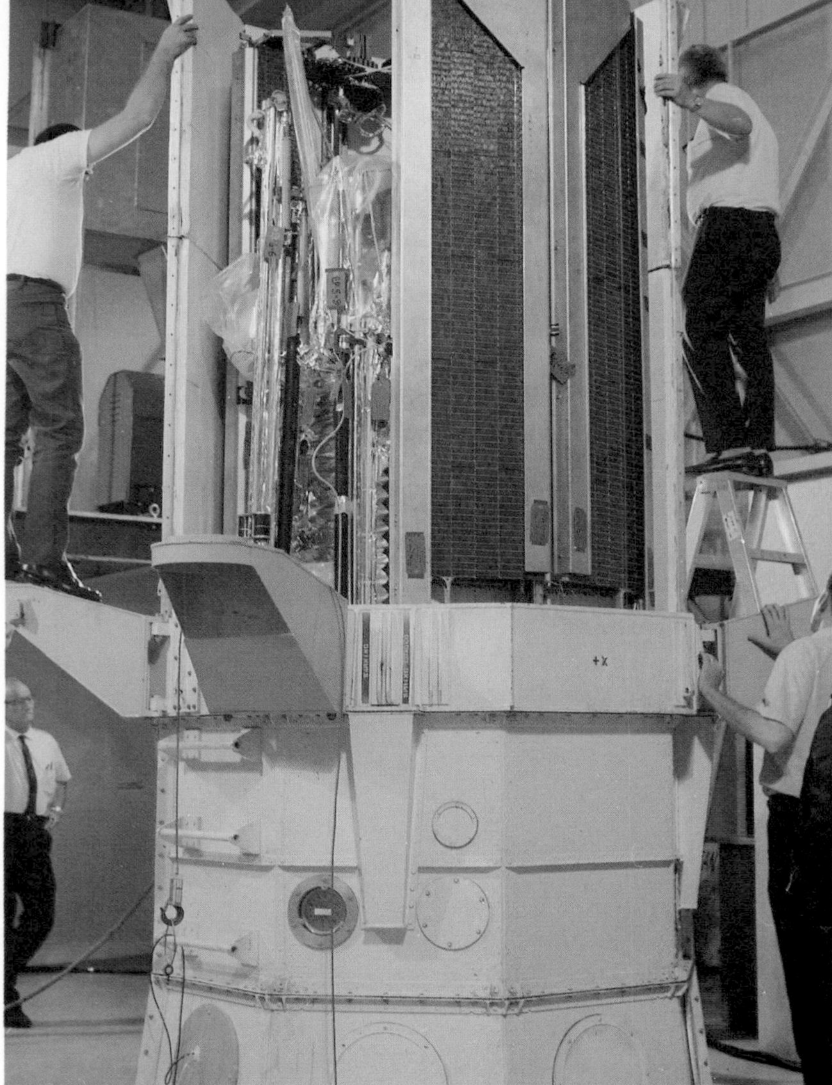

Above left: Orbiting Geophysical Observatory OGO 3 undergoes a flight readiness check at the Goddard Space Center. *Below left:* OGO 6, the last of the OGO craft — see the text on pages 52 and 56. *Above:* The Solar Maximum Mission SMM satellite, which made valuable discoveries about fluctuations in the sun. 'Solar Max', as this craft was dubbed, was launched on 14 February 1980, and its recovery by the Shuttle Orbiter Challenger in April 1984 is shown *at right.*

contributed to understanding of the chemistry of earth's atmosphere, of earth's magnetic field and of how solar particles penetrate and become trapped in the magnetosphere.

They provided the first evidence of a region of low energy electrons enveloping the high energy Van Allen Radiation Region, first observation of daylight auroras, first global map of airflow distribution, and much other knowledge about magnetic fields, particle radiation, earth's ionosphere, the shockwave between the geomagnetic field and solar wind that was discovered by Explorer 18, and the hydrogen cloud enveloping earth.

Beyond earth, they completed the first sky survey of hydrogen, discovered neutral hydrogen around the sun, found several strong sources of hydrogen in the Milky Way, and in April 1970, detected a cloud of hydrogen 10 times the size of the sun around comet Bennett. The existence of large amounts of hydrogen around comets was first discovered around comet Tago-Sato-Kosaka in January 1970 by Orbiting Astronomical Observatory 2 'Stargazer.' The large amounts of hydrogen around comets that were observed by OAO and OGO clearly establish that hydrogen is a major constituent of comets.

Biosatellite Experiments

The Biosatellite program was designed to study the effects of the condition popularly termed 'weightlessness' alone and in conjunction with a known source of radiation. The program was temporarily set back by failure of Biosatellite 1, launched 14 December 1965, when the craft could not be brought back to earth as planned because of retrorocket failure.

Biosatellite 2, launched 7 September 1967, carried microorganisms, frog eggs, plants and insects on an approximately 45-hour space flight. Among the wealth of scientific results harvested from this experiment were the following:

- Weightlessness appears to spur growth of wheat seedlings;
- Young, actively growing, rapidly dividing cells are more severely affected by radiation than mature, slowly growing, less actively dividing cells;
- Plant orientation is disturbed by weightlessness;
- While weightlessness appears to accelerate radiation-induced mutations and other cell damage, it also appears to slow the growth and metabolism of injured cells — the slower growth would provide more time to repair damage;
- Bacteria appear to multiply more rapidly in weightlessness than on earth — viruses failed, however, to reproduce as well as they do on earth;
- Gravity affects plants much more than previously realized — plant life on the orbiting Biosatellite 2 reacted more than animal life.

Biosatellite 3 was launched 28 June 1969, to gather data on how prolonged space flight affects bodily processes in primates. It contained a small monkey. Planned for 30 days, it was cut short on July 7 when the monkey became ill. After the capsule was recovered the monkey died. An autopsy showed death due to loss of body fluids. The experiment confirmed that replacement of body fluids that people lose during flight in space is vital to their well-being.

Biosatellite 3 was launched on 28 June 1969. It carried a highly instrumented female pigtail monkey named 'Bonny,' identical to the NASA laboratory monkey shown *below*. Biosatellite 3 is shown *at right*, during a preflight check at Cape Kennedy. Bonny fell sick during the flight, and died shortly after retrieval.

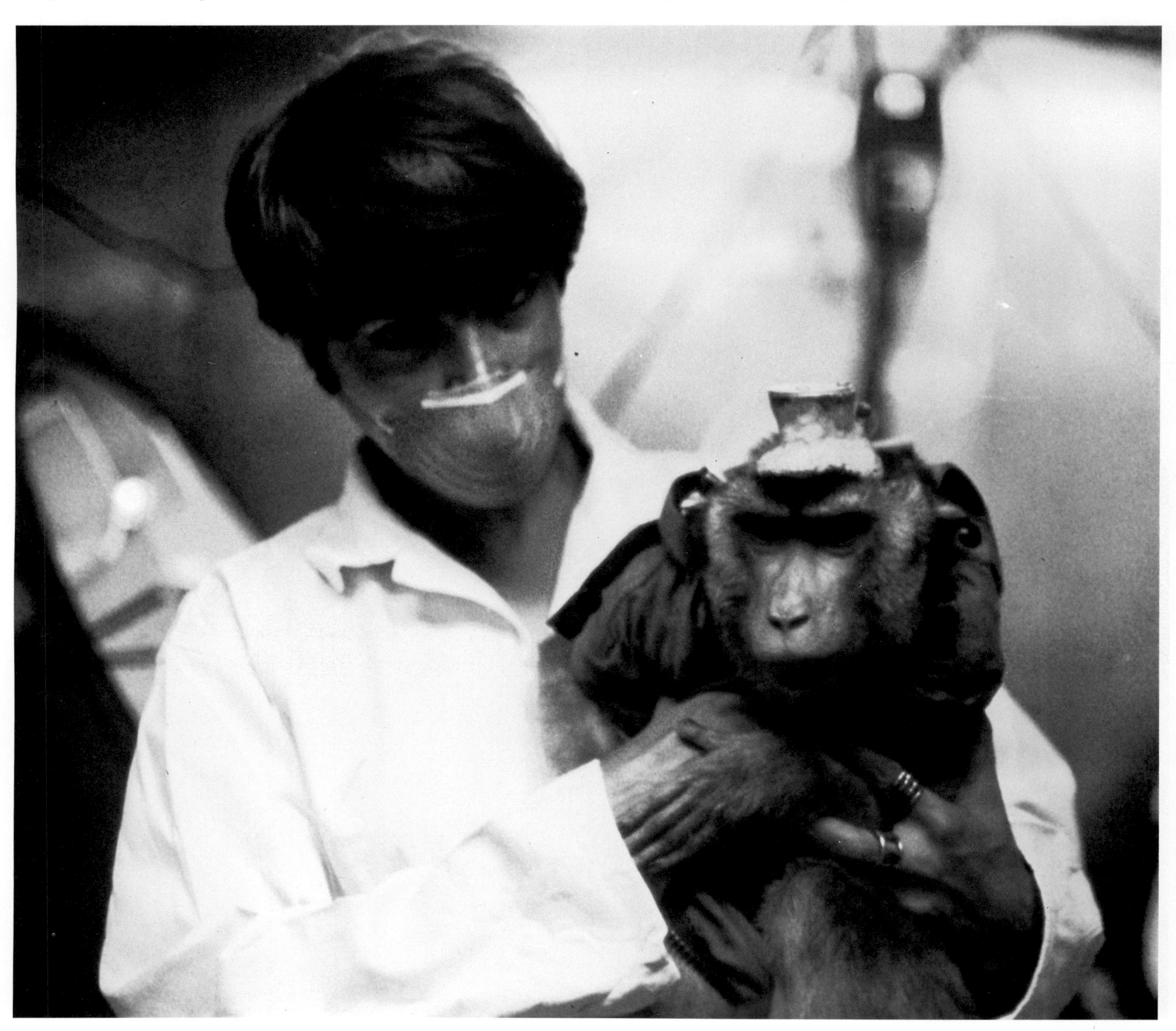

Solar Orbit Pioneers

Sun-orbiting Pioneers have contributed volumes of data about the solar wind, solar magnetic field, cosmic radiation, micrometeoroids, and other phenomena of interplanetary space. Pioneer 4, launched 11 March 1960, confirmed the existence of interplanetary magnetic fields and helped explain how solar flares trigger magnetic storms and the northern and southern lights (aurorae) on earth. The satellite also showed that the Forbush decrease of intergalactic cosmic rays near earth after a solar flare was the same in interplanetary space, and thus does not depend upon an earth-related phenomenon, such as the geomagnetic field.

Pioneers 6 through 9, launched 16 December 1965; 17 August 1966; 13 December 1967; and 8 November 1968, supplied volumes of data on the solar wind, magnetic and electrical fields, and cosmic rays in interplanetary space. Pioneer 7 detected effects of earth's magnetic field more than 3 million miles outward from the night side of earth. Pioneer 6 to 9 data also drew a new picture of the sun as the dominant phenomenon of interplanetary space. They found that the solar wind continues well beyond the orbit of Mars. (Pioneer 10, the first Jupiter explorer, continued to report its existence as it crossed the orbit of Pluto.) Analyses of their data indicated that the solar wind is an extension of the solar corona, the sun's atmosphere. They revealed that the wind draws out the sun's magnetic field to form what were previously called interplanetary magnetic fields but now is referred to as the heliosphere. They showed that the combination of the solar wind's outward pull and the rotation of the sun caused the lines of forces of the magnetic field to be twisted like streams of water from a whirling lawn sprinkler.

Pioneer data showed that solar cosmic rays spiral around the lines of force of the sun's magnetic field. This indicates they travel through space in well-defined streams.

Shown *below* during pre-flight preparations, Pioneer 5, launched 11 March 1960, established a deep space communication record for its time of 22.5 million miles. Pioneer 6 *(at right)* was launched on 16 December 1965, and Pioneer 8 *(at far right)* was launched on 13 December 1967. These returned data on the solar wind.

Other Space Science Activities

NASA also supports ground observatories. In 1976, ground observatories discovered a satellite circling Pluto and ascertained that methane ice covered the planet's surface. With these observations, astronomers were able to increase their accuracy in measuring Pluto's mass. Their calculations resulted in a substantial reduction of estimates of Pluto's mass and indirectly increased support for the theory that a massive solar system object existed beyond Pluto. The reason is that Pluto's mass, as currently estimated, is far less than adequate to produce the perturbations in the orbits of Uranus and Neptune that inadvertently led to the search for and discovery of Pluto.

In 1982, ground observatories reported that Pluto has an atmosphere of gaseous methane. The atmosphere may be the result of Pluto being closer to the sun as its elliptical orbit has carried it inside of the orbit of Neptune. In 1983, ground observatories discovered a quasar 12 billion light years away, the most distant object discovered. The search that resulted in this discovery started 10 years ago, using antennas of NASA's deep space tracking and data acquistion network and observatories in the United Kingdom and Australia.

NASA's Project SETI, Search for Extraterrestrial Intelligence, was begun in 1960. It not only searches for radio signals that could be from intelligent creatures, but also does radio astronomy mapping of the sky and studies manmade radio interference that could affect space tracking, data acquisition, and communication. Helios 1 and 2, a cooperative project of NASA and the Federal Republic of Germany, were designed to survive and function at distances closer to the sun than any other spacecraft.

From perihelions lower than 45 million kilometers (28 million miles) on opposite sides of the sun, they added to knowledge about the solar corona, magnetic field, wind and radiation and about micrometeroids and other phenomena in this vicinity of space. Helios is named for the sun god of ancient Greece. The craft were launched on 10 December 1974 and 15 January 1976.

NASA Deep Space Network outposts like Rosman Station *(below)* gather data from satellites and other spacecraft, so that it can be deciphered into usable form. Perhaps a yet-nonexistent NASA space probe will one day return photographic data to match the artist's conception *at right*—of Pluto and its satellite Charon.

Tracking and Data Acquisition

by Jim Kukowski

In the late 1950s numerous problems confronted the fledgling space effort. Not only was it difficult to place a satellite into earth orbit, it was equally difficult to locate and track a satellite once orbit was achieved.

Ground observations of satellites were relied upon heavily to gain orbital information. Data acquisition was an even more difficult task.

Sounding rocket flights to high altitudes in the 1950s paved the way for the development of the early tracking network that was established by the United States. Telescopes and radars followed the short flights of the rockets loaded with 'quick-look' scientific instruments. It was necessary to improve and expand that capability to track and acquire information from the earth satellite.

NASA's Space Tracking and Data Acquisition Network (STDN) evolved from the Minitrack tracking stations (11 of them) set up by the US Naval Research Laboratory for the Vanguard Program in 1956 and 1957. In the early days of the space effort, Minitrack was the main method for tracking. Its radio interferometers formed electronic 'fences' to search the sky for any spacecraft carrying 136 megahertz radio beacons.

Operating concurrently with Minitrack was the Smithsonian Astrophysical Observatory (SAO) Network. SAO telescope sites located around the world provided valuable information to the growing tracking needs of the United States space effort.

As the STDN network became more sophisticated and advances were made in data acquisition techniques, the network grew in size and scope. Great strides were made in the collection of data from the satellites through the development of large steerable antennas.

The Goddard Space Flight Center in Greenbelt, Maryland, was the hub of the growing network and remains so today. In addition to managing the growing number of tracking stations, Goddard was also the central facility of the worldwide NASA Communications Network (NASCOM).

With the decision to develop a spacecraft program to take man into space, NASA took another significant step by developing a sophisticated tracking and data acquisition system using advanced techniques in radar.

A worldwide radar, telemetry and communications network was devised to support the manned flight program. The 15-station Mercury Network became operational on 1 July 1961 and performed superbly during the Mercury Program. More advances were made for the follow-on Gemini flights in the mid 1960s and the name of the network was changed to the Manned Space Flight Network (MSFN).

New requirements added to the sophistication of the STDN network during the 1960s and early 1970s. Data flow from a variety of scientific, communications, meteorological and earth resources satellites increased in amount and complexity. Of particular significance was the imaging from the Earth Resources Technology Satellite program, later to be known as Landsat. So valuable were the pictures from space, provided by the spacecraft, that a dozen nations would eventually construct their own tracking stations to receive the images.

With the moon as the eventual target for the manned program it was necessary to advance the state of the art for these lunar landing missions. A new name in tracking and data acquisition came to the forefront . . . Unified S-Band.

This was the unification of all tracking and communications functions (voice, telemetry and command) into a single communications link. MSFN was comprised of the ground stations, ships at sea and antennae carrying aircraft all linked together by the globe spanning NASCOM communications network. The system provided the highly technical data flow including the stunning television transmissions from the surface of the moon.

Even as the lunar program was going on, network planners were devising the next step . . . the Tracking and Data Relay Satellite System (TDRSS). TDRSS was the answer to the requirement for nearly continuous communications with newer and more sophisticated satellites.

Instead of the existing worldwide network of ground stations which can provide coverage up to only 15 to 20 percent of each orbit, TDRSS, consisting of two satellites, an in-orbit spare and a single ground station, will provide nearly full-time coverage not only for the operational Space Shuttle, but also for as many as 26 other earth-orbiting satellites simultaneously.

At top right: **An artist's conception of a Tracking and Data Relay Satellite. The first TDRS was lost with Space Shuttle *Challenger* in the 51-L (STS-25) tragedy; *Discovery* STS-25, however, launched a TDRS on 29 September 1988 (see also page 181). *At right and opposite:* Tracking and Data Relay Satellite System ground stations.**

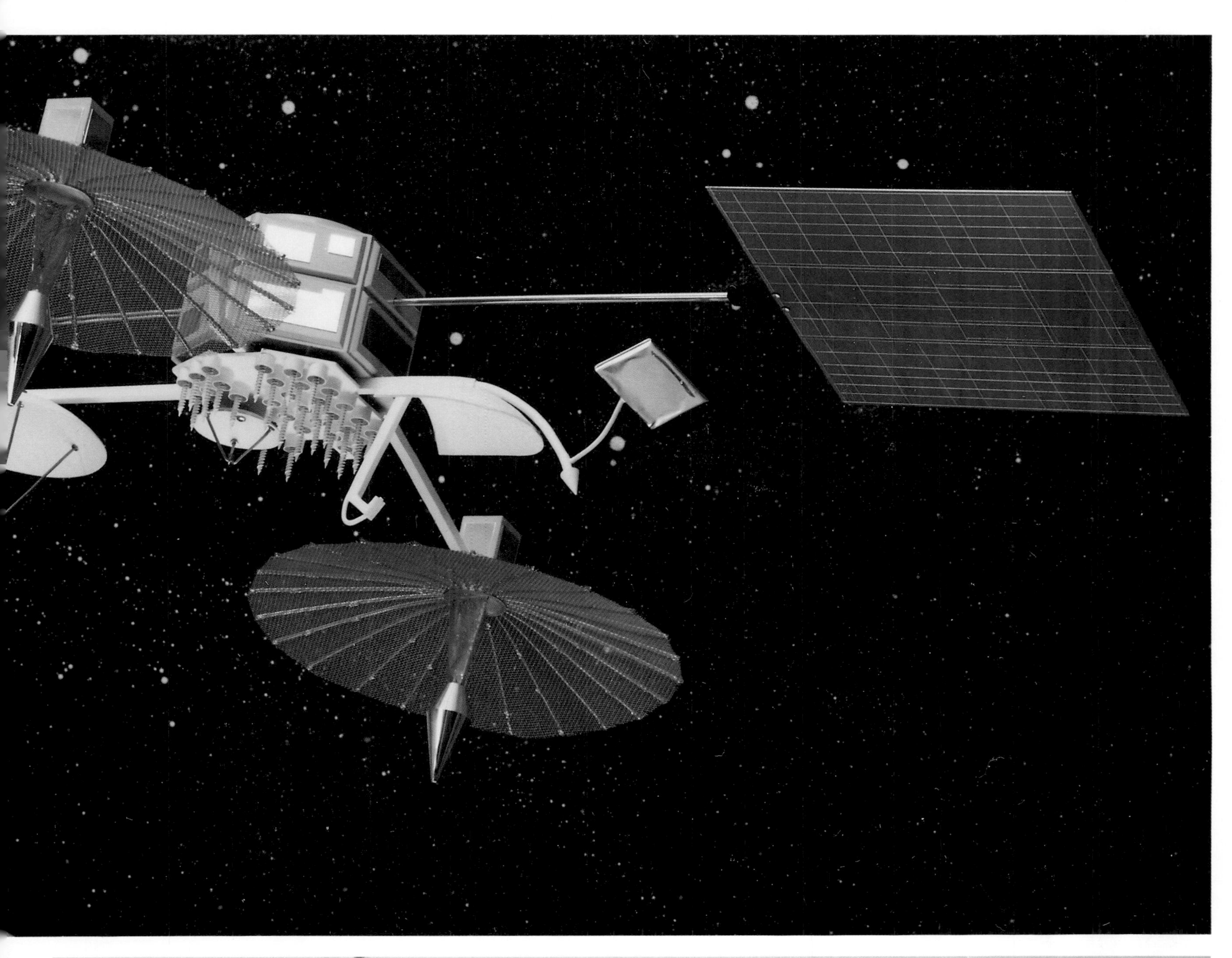

INTERPLANETARY SPACECRAFT

By the end of 1989, NASA's fleet of interplanetary travellers will have conducted close-up scientific and photographic observations of every planet in the solar system except Pluto. In a little over a quarter century our knowledge of our planetary neighbors has increased a thousandfold over what was learned in all preceding recorded time. Mariner 2, which confirmed that beneath its bright cloud cover, Venus was a dry lifeless inferno, also confirmed the existence of the solar wind as a predominant feature of interplanetary space. Pioneer 10, which provided the first close-up of Jupiter, returned data indicating that the Gegenschein — a faint glare in earth's sky directly opposite the sun — and the zodiacal light are due to the sunlight reflecting from small particles in interplanetary space rather than in earth's atmosphere. The zodiacal light is a faint cone of light extending upward from the horizon in the direction of the zodiac, or ecliptic. Pioneer 10 also showed that the heliosphere extends beyond Jupiter.

Mariner Spacecraft

Mariner 2 was America's first successful planetary spacecraft. Launched 27 August 1962, it flew past and made close-up observations of Venus on 14 December 1962. Its data supported earth-based microwave scans that suggested a surface temperature as high as 800 Fahrenheit, hot enough to melt lead on both day and night sides. It detected no openings in the dense clouds enveloping Venus. Its data indicated no intrinsic Venusian magnetic field nor increase in radiation. This suggested that Venus has no radiation belt like the Van Allen Radiation Region around earth. The data are consistent because the Van Allen Radiation Region is attributed to the capture of energetic particles by earth's magnetic field. Mariner 2 also confirmed the predominance of the solar wind as a feature of interplanetary space and the ubiquity of interplanetary magnetic fields which, scientists now realize, are an extension of the sun's magnetic field dragged out into space by the solar wind.

Speeding by Mars on 14 July 1965, Mariner 4, launched 28 November 1964, gave the world its first close look at that planet's surface. The pictures were surprising: a heavily cratered moon-like surface that looked like it may not have changed much in billions of years. Because the pictures covered about one percent of Mars, they permitted no conclusions until other spacecraft viewed additional areas of the planet. The pictures covered some areas crossed by the supposed Martian canals but showed no readily apparent straight-line features that could be interpreted as artificial. Among other data from Mariner 4 were indications that the surface atmospheric pressure on Mars was less than 10 millibars. Earth's sea level air pressure is about 1000 millibars. Humans would need a pressure suit on Mars. Mariner 4 also gave additional information about Mars, size, gravity and path around the sun. Mariner 4 detected neither a Martian magnetic field nor radiation belt but revealed a Martian ionosphere.

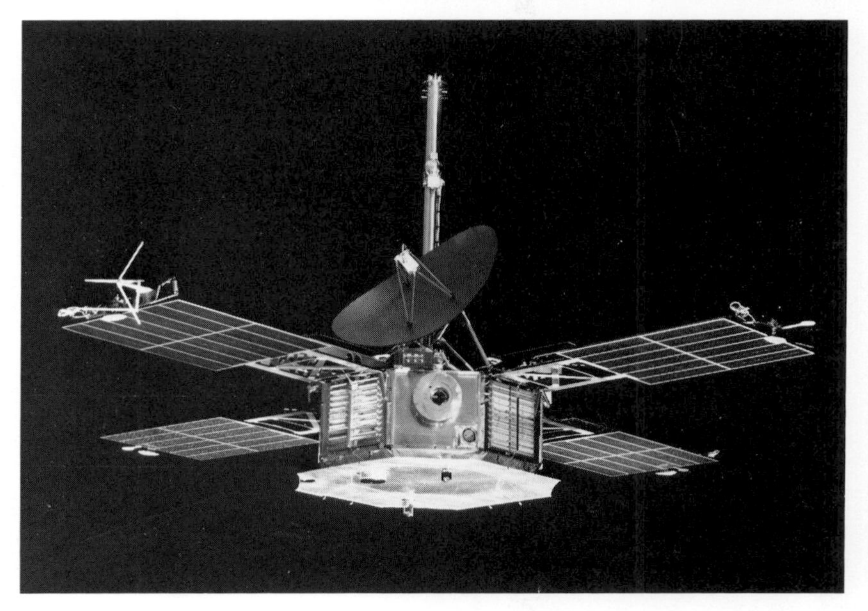

Mariner 1 (shown *at right*, in an artist's conception) was destroyed in a launch failure on 22 July 1962. Its twin, Mariner 2 (a display model of which is shown *at top, above*), was successfully launched toward Venus on 27 August 1962. Mariner 4 *(above, center)*, launched on 28 November 1964, returned man's first closeup photos of Mars. *Above:* The Venus atmospheric probe Mariner 5, launched 14 June 1967.

Mariner 5 was launched 14 June 1967 to refine and supplement data about Venus obtained from Mariner 2 and other observations. It contained improved instrumentation and, in October 1967, flew within 2500 miles of Venus as compared to the 21,645-mile closest approach of Mariner 2. Among conclusions drawn after this flyby were:

- Venus' atmosphere is at least 80 percent carbon dioxide.
- Venus' atmosphere is about 100 times denser than earth's.
- Venus' surface temperature may be as high as 800 F.
- The solar wind is diverted around Venus by the planet's ionosphere. (Earth's magnetic field diverts the solar wind around our planet.)
- The Venusian exosphere, like earth's, is made up largely of hydrogen.
- Venus has no detectable magnetic field nor radiation belt.

Mariners 6 and 7 were launched on 25 February and 27 March 1969,

respectively, and flew as close as 2000 miles to Mars on 31 July and 5 August 1969, respectively. Mariner 6 flew past the planet along its equator. Mariner 7 overlapped part of the Mariner 6 ground track and then sped south over the south polar ice cap. They took pictures of Mars and studied it with infrared and ultraviolet sensors. Their pictures show not only cratered but also smooth and chaotic surfaces.

The chaotic region, about the size of Alaska, is characterized by short ridges, slumped valleys, and other irregularities that resemble the after effects of a landslide or quake. Nowhere on earth is a comparable feature so vast.

Launched 30 May 1971, Mariner 9 went into Martian orbit on 13 November 1971, the first spacecraft placed into orbit around another planet. It orbited and studied Mars and the planet's two tiny satellites, Deimos and Phobos, until 27 October 1972. Mariner 9 arrived at a discouraging time when a dust storm enveloped most of the planet. Even the dust storm, however, provided information of value such as atmospheric circulation pattern and the fact that only on Mars were dust storms of such magnitude observed. When the storm cleared, Mariner 9 was able to photograph Martian geography in remarkable detail. Its photographs show Martian volcanic mountains, such as 15.5-mile-high Olympus Mons, which is larger than any mountain on earth. Also discovered was the vast 3000-mile Valles Marinaris (Mariner Valley), long enough to stretch across the United States from the Atlantic to the Pacific Ocean; and signs that rivers and possibly seas may have existed on Mars.

Mars' two small satellites, Deimos and Phobos, appear as points of light in ground observatory telescopes. Mariner 9 swept close to both, providing pictures that showed them to be irregularly shaped and heavily cratered.

Mariner 6, shown *below*, took 75 photos of Mars during its July 1969 encounter. *At right:* A 27 March 1969 launch photo of Mariner 7, which was to fly over Mars' north pole. Mars probe Mariner 9 *(at far right)* became the first earth spacecraft to be placed in orbit around another planet, on 13 November 1971. Mariner 10 *(below opposite)* encountered Venus once, and Mercury three times, in 1974–75.

Mariner 10, launched 3 November 1973, flew by Venus on 5 February 1974 and in a solar orbit, swept nearby and gathered information about Mercury on three separate occasions: 29 March and 21 September 1974 and 16 March 1975. These first close-ups of Mercury reveal an ancient surface bearing the scars of huge meteorites that crashed into it billions of years ago. They show unique large scarps (cliffs) that appear to have been caused by crustal compression when the planet's interior cooled. A Mercurian magnetic field, about a hundredth the magnitude of earth's, and an atmosphere, about a trillionth the density of earth's, were detected. The Mercurian atmosphere is made up of argon, neon and helium. Data suggest a heavy iron-rich core making up about half the planet's volume. (Earth's core is about 25 percent of its volume.) Mariner 10 reported that Mercury's surface temperatures were 950° F on the sunlit side and -350° F on the night side.

Mariner 10 was the first picture-taking spacecraft to view Venus. Its optical cameras however failed to find an opening in the clouds that shroud the planet. The spacecraft's ultraviolet cameras revealed that Venus' topmost clouds circled the planet 60 times faster than Venus rotates.

Mariner 10 also confirmed a long-held theory about earth's weather that solar heat causes air to rise in the tropical area, flow to the poles, cool, fall and then return to the tropics where the process is repeated. No such process could be discerned on earth as earth's rapid rotation, variable atmospheric water content, sizable axial tilt and mixing of continental and ocean air masses produce strong air currents obscuring the equatorial—polar flow. Venus has practically no water vapor, rotates slowly, has no axial tilt, and no mixing of continental and ocean air masses (no oceans) to obscure this flow.

***Below:* A schematic drawing of Mariner 10. Launched on 3 November 1973, Mariner 10 was the first and only Mercury probe; and it was the first to use one planet's gravitational field to 'whip' it toward another; the flight path is illustrated *below, at bottom. At right:* A mockup of one of Mariner 10's three Mercury flybys.**

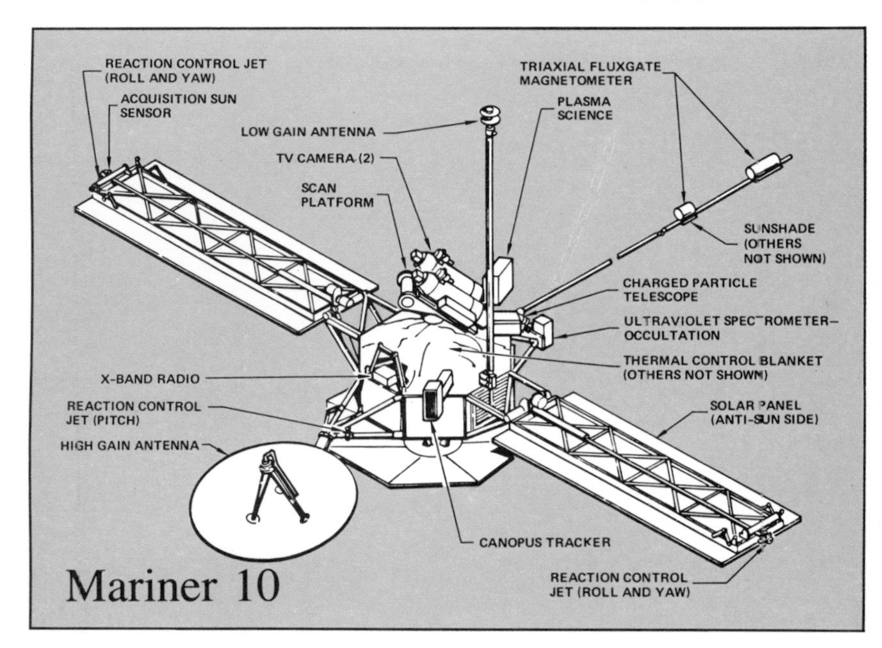

The Deep Space Pioneers

Pioneer 10, launched 3 March 1972, was the first spacecraft to cross the Asteroid Belt, first to make close range observations of the Jupiter system, and the first to go beyond the outermost planets. In December 1973, it swept nearby Jupiter, finding no solid surface under the thick and deep clouds enveloping the planet. Thus, the world learned that Jupiter is a planet of liquid hydrogen. It explored the huge Jupiter magnetosphere, made close-up pictures of the Great Red Spot and other atmospheric features, and observed and measured at relatively close range Jupiter's large Galilean satellites — Io, Europa, Ganymede and Callisto.

Passing Jupiter, Pioneer 10 continued to map the heliosphere, the giant solar magnetic field drawn out from the sun by the solar wind. Pioneer 10 found that the heliosphere, like the magnetospheres of earth and Jupiter, behaves like a cosmic jellyfish, altering its shape in response to rises and falls in solar activity. It also reported that the speed of the solar wind does not decrease with distance from the sun. On 13 June 1983, Pioneer 10 crossed the orbit of Neptune which will be the planet farthest from the sun until the turn of the century. (This is because Pluto, although farthest on average, has an extremely elliptical orbit that crosses and goes inside of Neptune's.)

Pioneer 10 is searching for the limits, or outer boundary, of the heliosphere. Together with Pioneer 11, it is also searching for a mysterious massive object beyond the known planets. Scientists hypothesize the existence of the object because of unexplained irregularities in the orbits of Uranus and Neptune. Pluto's mass is insufficient to cause these irregularities.

Considering the possibility, however remote, that Pioneer 10 may encounter intelligent extraterrestrials, NASA equipped Pioneer 10 with a plaque. The plaque has diagrams, sketches and binary numbers indicating where, when and by whom Pioneer 10 was launched.

Launched 6 April 1973, Pioneer 11 (a sister ship to Pioneer 10) swept as close to Jupiter as 42,000 kilometers, compared to the 129,000-km closest approach of Pioneer 10, on 4 December 1973. It provided additional detailed data and pictures on Jupiter and its satellites, including the first look at Jupiter's north and south poles which cannot be seen from earth. This view was possible as Pioneer 11 was guided so that Jovian gravity actually threw the craft out of the plane of the ecliptic in which the planets lie. From above this plane, Pioneer 11 was able to confirm that the heliosphere extends outward in all directions from the sun and is broken into northern and southern hemispheres by a bobbing sheet of electric current.

Pioneer 11 passed nearby Saturn on 1 September 1979, demonstrating a safe flight path for the more sophisticated Voyagers spacecraft that NASA would launch in 1977. It provided the first close-up observations of Saturn, its rings, satellites, magnetic field, radiation belts and stormy atmosphere. It showed areas, smaller but resembling the Great Red Spot on Jupiter, in Saturn's clouds.

Pioneer 11 found no solid surface on Saturn but discovered at least one additional satellite and ring. Its data suggested that Saturn's rings are icy in composition with little or no rock or metal and that Saturn's three largest satellites — Titan, Rhea and Iapetus — are composed in large part of ice.

Launched in 1972, the first outer Solar System probe, Pioneer 10 *(below)*, studied the heliosphere and the planet Jupiter. Pioneer 11 *(at right)* used 'gravity assist' (see text, above, and caption, page 68) in its Jupiter and Saturn encounters.

Viking on Mars

A pair of two-part Viking spacecraft made the world's most extensive study of Mars. The project used two spacecraft, each of which had an orbiter and lander. Viking 1, launched 20 August 1975, went into Martian orbit on 19 July 1976, and put its lander on the surface 20 July 1976. The orbiter stopped transmitting 7 August 1980; the lander, late in November 1982.

Viking 2, launched 9 September 1975, went into Martian orbit on 7 August 1976. Its lander touched down on 3 September 1976. The orbiter reported until 24 July 1975; the lander, 12 April 1980. They returned a wealth of photographs and other data, mapping about 97 percent of Mars. Their success generated so much public enthusiasm that about $50,000 was raised in 1980 to prolong the project. Late in 1980, the Viking 1 lander was renamed Mutch Memorial Station in memory of Dr Thomas A Mutch, former Viking Lander Imaging Team leader and former NASA Associate Administrator for Space Science who had disappeared while mountain-climbing in the Himalayas.

Among the significant Viking discoveries:

- The Martian atmosphere, although too thin for most living things on earth to survive (about a 1/100 as dense as earth's), contains all components necessary to sustain life: nitrogen, carbon, oxygen and water vapor.
- The Martian surface resembles deserts on earth but is drier than earth's driest desert. However, considerable quantities of water are locked in the north polar ice cap and in the form of subsurface permafrost.
- Mars' northern and southern hemispheres are very different climatically. Global dust storms originate in the southern hemisphere while water vapor is comparatively abundant only in the far north during its summer.
- Lander analyses of Martian soil gave puzzling results that neither proved nor disproved the presence of past or present life. However, their failure to detect organic molecules reduced support for the presence of life forms.
- No canals or artifacts of any kind were found on Mars. However, evidence was found that in the past Mars' atmosphere was much thicker, its temperature was warmer and it had water on its surface.
- While volcanic mountains — at least one of which is bigger than any on earth — were found on Mars, the planet is seismically much less active than earth.

Viking Lander 1 is shown *below*, with its sterile capsule in place, in the Assembly Building at Kennedy Space Center. *At far left*, a Viking lander and Orbiter as mated, and *at left*, with solar panels folded, awaiting the emplacement of their launch shroud. *Below opposite:* The Viking 2 launch on 9 September 1975.

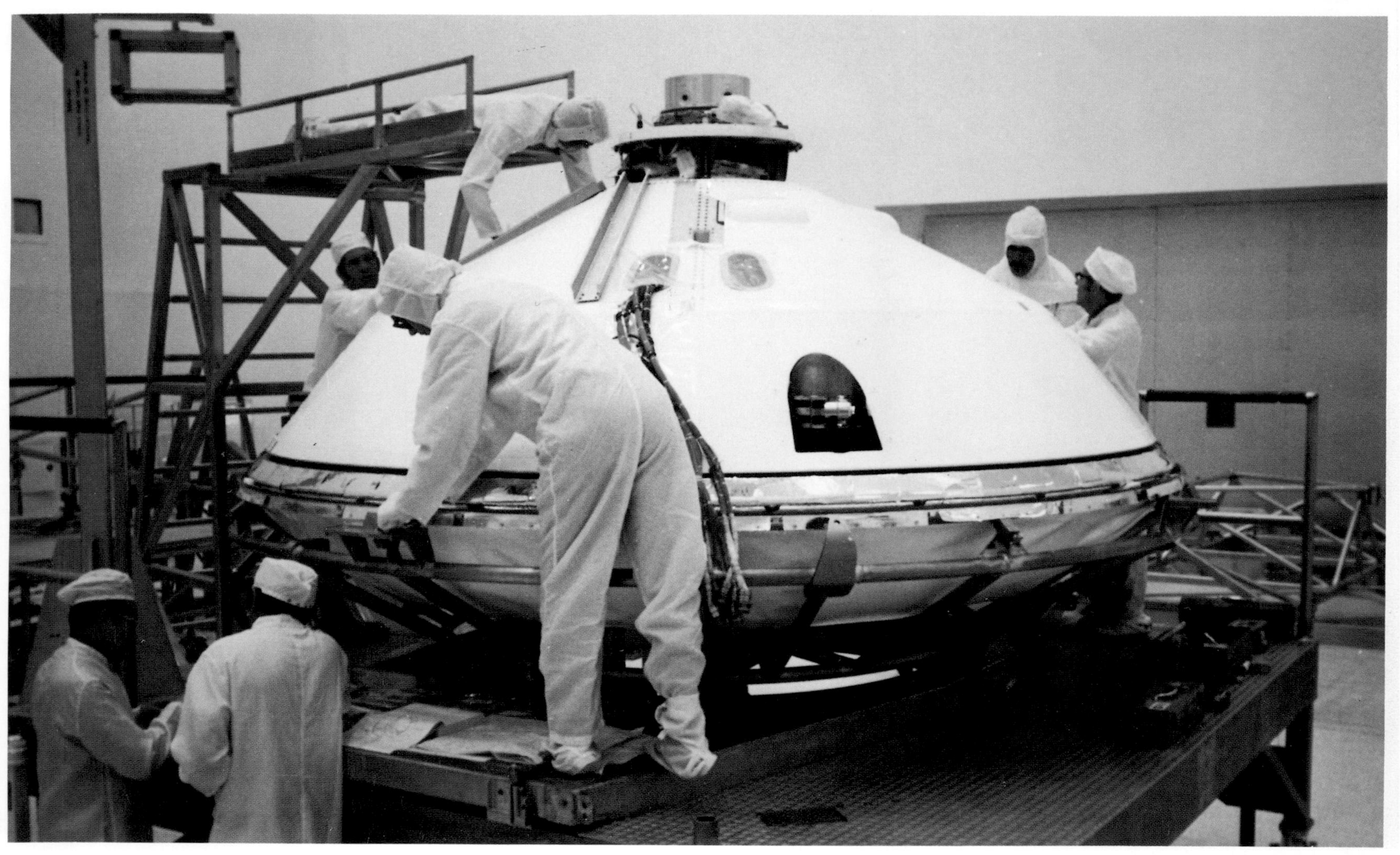

Above: Viking Lander 1 took this photo of the surface of Mars on 8 August 1978, during the 730th Mars day after its landing. *Below* is mockup of a Viking lander on Mars, complete with nonexistent white Mars sand. *At right* is an oblique view across Mars' Argyre Planitia, the closest of the large impact basins seen here. This view was obtained on 11 July 1976 by Viking Orbiter 1. Also evident in this photo is a carbon dioxide/water crystal haze, 15 to 25 miles above Mars' surface.

The Fabulous Voyager Program

Just as the solar system has an entourage of planets and other bodies, the giant planets Jupiter and Saturn have their own entourages of rings and satellites. NASA's sophisticated Voyagers 1 and 2 provided a wealth of information including discoveries and answers to questions that have eluded astronomers since the birth of civilization.

Voyager 2 was launched on 20 August 1977 and followed by Voyager 1 on 5 September. Because of its trajectory, Voyager 1 reached Jupiter first — on 5 March 1979 — and its identical sister ship arrived in early July. After leaving Jupiter, Voyager 1 continued to pull ahead of its partner, reaching Saturn in November 1980, while Voyager 2 flew by Saturn in August 1981. After passing Saturn, Voyager 1 was commanded to turn north and climb up away from the ecliptic plane where the planets orbit, so as to obtain close-up photos of Saturn's mysterious moon Titan. Voyager 2 continued outward across the ecliptic toward a January 1986 flyby of Uranus and a September 1989 rendezvous with Neptune.

Voyager's discoveries were numerous. They included: a ring of rocky debris around Jupiter, an ionized torus of sulphur around Jupiter, vast thunderstorms, sometimes big enough to engulf earth, on Jupiter and Saturn.

In addition, it discovered many additional small satellites, the natures of the Galilean satellites and Saturn's largest satellites, discoveries about Saturn's rings such as structure, width, unusual braiding, spoke patterns, ringlet formations, additional rings and numerous ringlets filling most of what appear from earth to be gaps in Saturn's rings.

Voyager's close looks at the Galilean satellites showed that gaudy, sulphurous Io is the most volcanic object in the solar system; Europa has the smoothest surface; Ganymede had some tectonic activity before it froze solid about 3 million years ago; and Callisto's crater-pocked surface is as ancient as that of the moon or Mercury. Ganymede and Callisto are composed of about equal parts of water ice and rock; Europa has substantial quantities of water; while Io is a waterless moon.

Voyager's findings indicate that Saturn's rings and satellites are for the most part composed of water ice. The exceptions are the giant satellite Titan which is half rock and the outermost satellite Phoebe which is mostly rock. Voyager photographs of Phoebe show that it resembles asteroids. It may be a captured outer solar system asteroid.

Voyager measurements toppled Titan from its position as the solar system's largest satellite. They found Jupiter's Ganymede to be slightly larger. Both are bigger than either of the planets Mercury or Pluto.

Voyager 1 discovered that nitrogen makes up about 82 percent of Titan's atmosphere. (Nitrogen is 78 percent of earth's atmosphere.) The other known ingredients of Titan's atmosphere are methane, ethane, acetylene, ethylene, hydrogen, cyanide and other organic compounds. Titan's atmosphere is believed to be similar to earth's in primeval times. Titan's composition reflects an abundance of water. An orange-colored smog, thought to result from chemical reaction of solar radiation on the methane in Titan's atmosphere, envelops the satellite, preventing direct observation of its surface. Voyager found Titan's surface air pressure to be 1.6 times that of earth's sea level pressure. Some scientists speculated that the smog in Titan's atmosphere could have created a greenhouse effect that over the last few billion years warmed the planet enough for organic chemicals to evolve toward prelife forms. Voyager 2 measurements show that Titan's surface temperature is 95 Kelvin (-288° F), too cold for water to liquefy or for significant progress in prelife chemistry.

Around this temperature, methane could exist as a liquid, vapor or solid, just as water does on earth. On Titan then, methane rain or snow may fall

Below, at bottom: **An explanatory diagram of a Voyager spacecraft.** ***Below:*** **Technicians prepare a 'Sounds of Earth' recording and an American flag for placement aboard Voyager 2.** ***At right:*** **An artist's conception of a Voyager flyby of Saturn.**

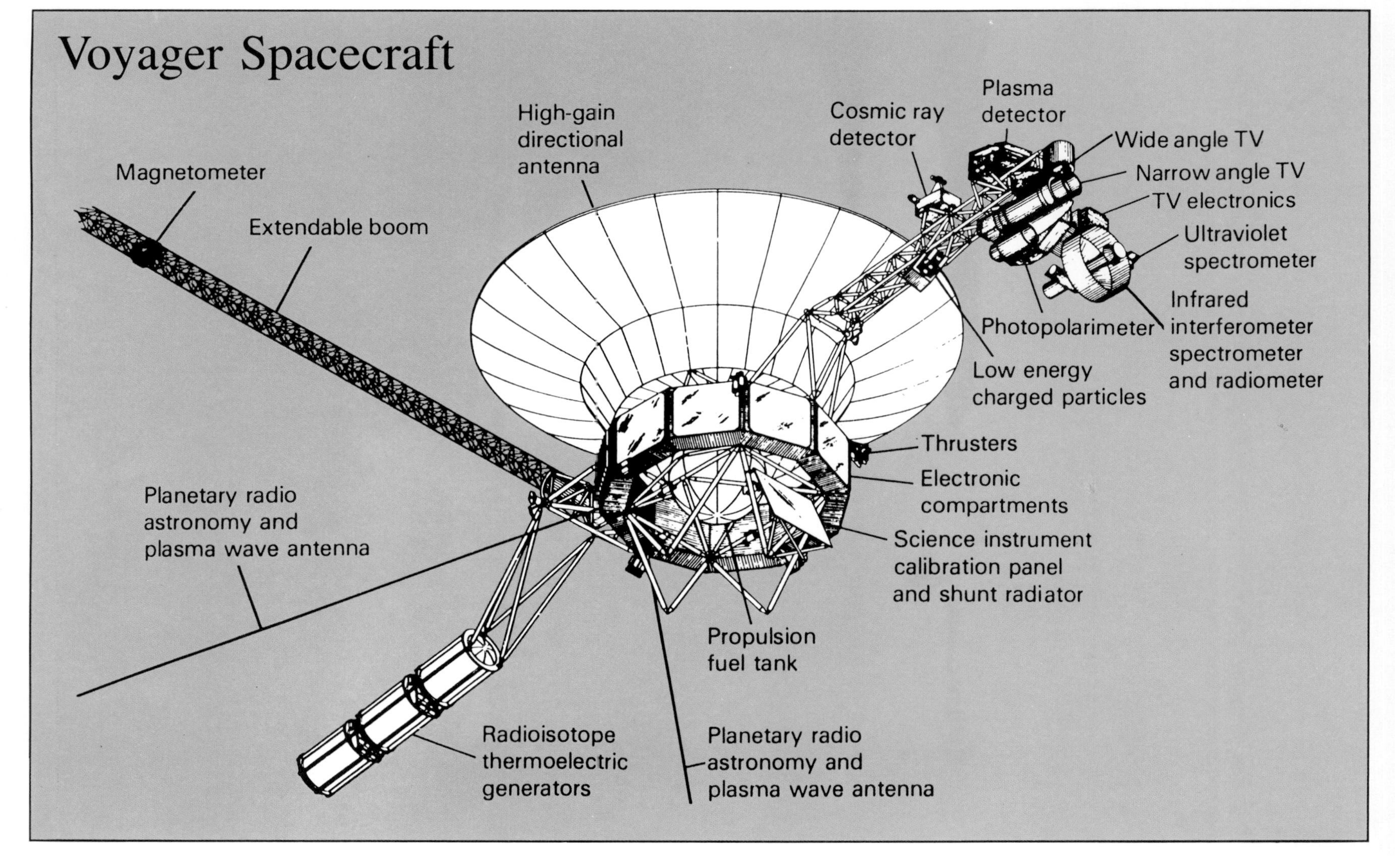

from methane clouds. Methane rivers may flow through icy methane channels and methane oceans may fill icy basins.

The Voyagers also confirmed that neither Jupiter nor Saturn have solid surfaces. Their data indicated that while Jupiter's radiant heat may originate from gravitational contraction or release of heat accumulated during its formation, Saturn's comes from gravitational separation of helium and hydrogen in Saturn's interior. Both planets radiate about 2½ times more heat than they receive from the sun.

A rare planetary alignment when Voyager 2 was launched will enable it to sweep near Uranus in January 1986 and Neptune in August 1980. If still functioning, it will provide the world with its first close-ups of these planets.

Voyagers 1 and 2 will eventually leave our solar system, like Pioneer 10. Each carries a record called 'Sounds of Earth' with electronically imprinted words, photographs and diagrams telling about earth. The 'Sounds' include greetings in sixty languages, music from different cultures and eras and sounds of the wind, surf, animals and other earth phenomena.

Pioneer-Venus

The Pioneer-Venus mission included an orbiter, launched 20 May 1978 and placed into Venusian orbit on 4 December 1978, as well as a multiprobe bus, launched 8 August 1978, that separated about three weeks before entry into Venus' atmosphere into four probes and the bus. The five entered Venus' atmosphere at widely separated locations on 9 December 1978 and returned information as they descended to the surface. Although none was designed to survive landing, one probe transmitted data for an hour afterward.

Orbiter radar data provided a topographic map of about 90 percent of Venus. Most of Venus is a gently rolling plain. There are two prominent plateaus: one as large as Australia, the other as large as the upper half of Africa. There is a volcanic structure larger than earth's Hawaii-Midway chain — earth's largest — and a mountain that towers higher over Venus' great plain than earth's Mount Everest over sea level.

Other data indicated two major volcanic areas that vent the planet's internal heat. This makes Venus the third solar system body — the others are earth and the Jupiter satellite Io — with significant volcanic activity.

Orbiter and probe data refined information about Venus' atmosphere. They showed that the temperature at Venus' surface is 900° F and air pressure on Venus is about 100 times greater than earth's sea level pressure.

The composition of Venus' lower atmosphere is 96 percent carbon dioxide, 3 percent nitrogen and 1 percent other gases, including extremely small parts of sulphur dioxide and water vapor. Venus' clouds are composed of three distinct layers, all of which consist mostly of corrosive sulphuric acid droplets.

Pioneer-Venus confirmed that the greenhouse effect is responsible for Venus' inferno-like surface temperatures; it also supported information

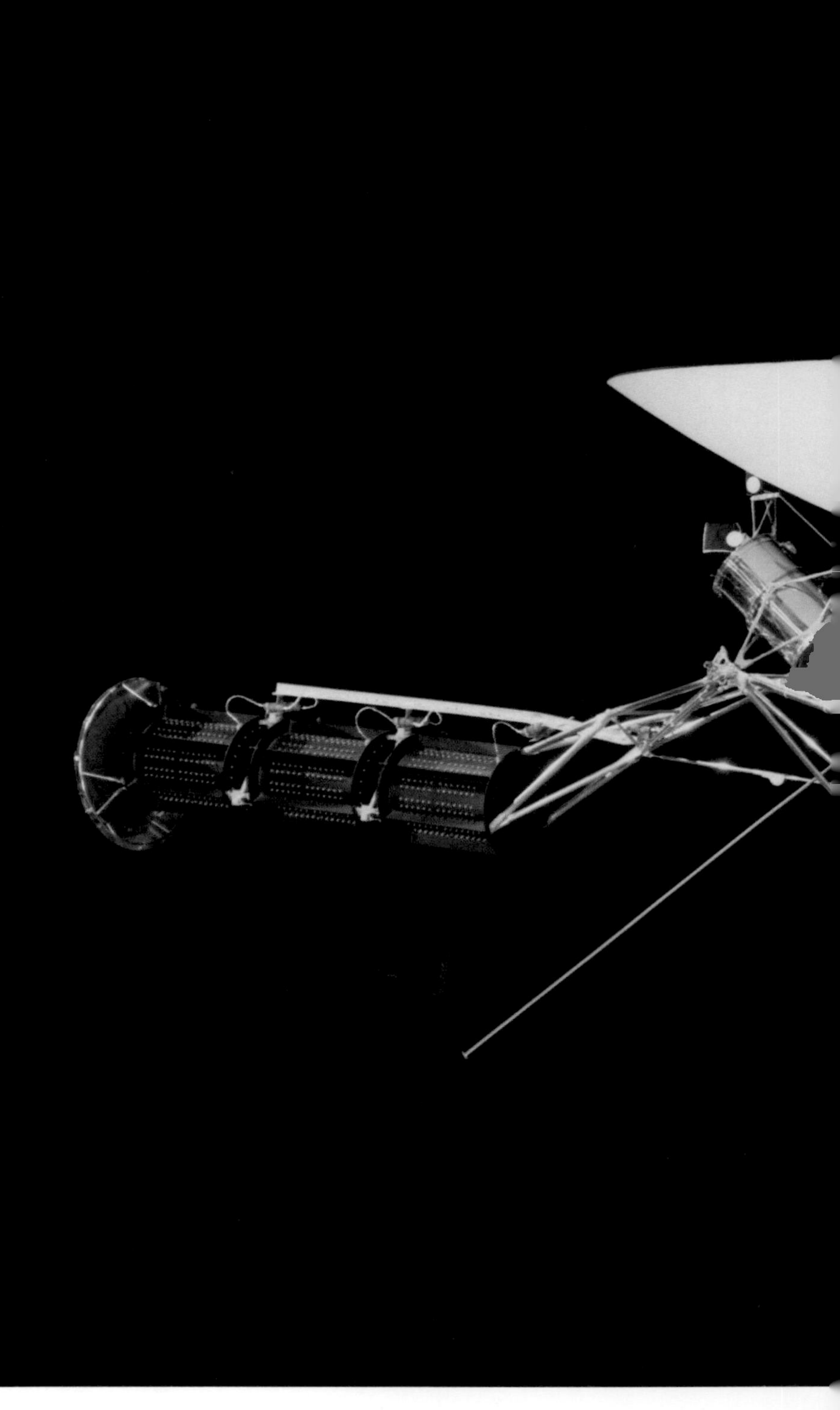

Voyager 2 encountered Jupiter in the spring of 1979, and sent back information that enabled NASA to make this cylindrical projection *(at top right)* of the giant planet; Voyager 2 went on to encounter Uranus on 24 January 1986, and returned such as the color photo *below* of that planet. *At right:* A Voyager spacecraft.

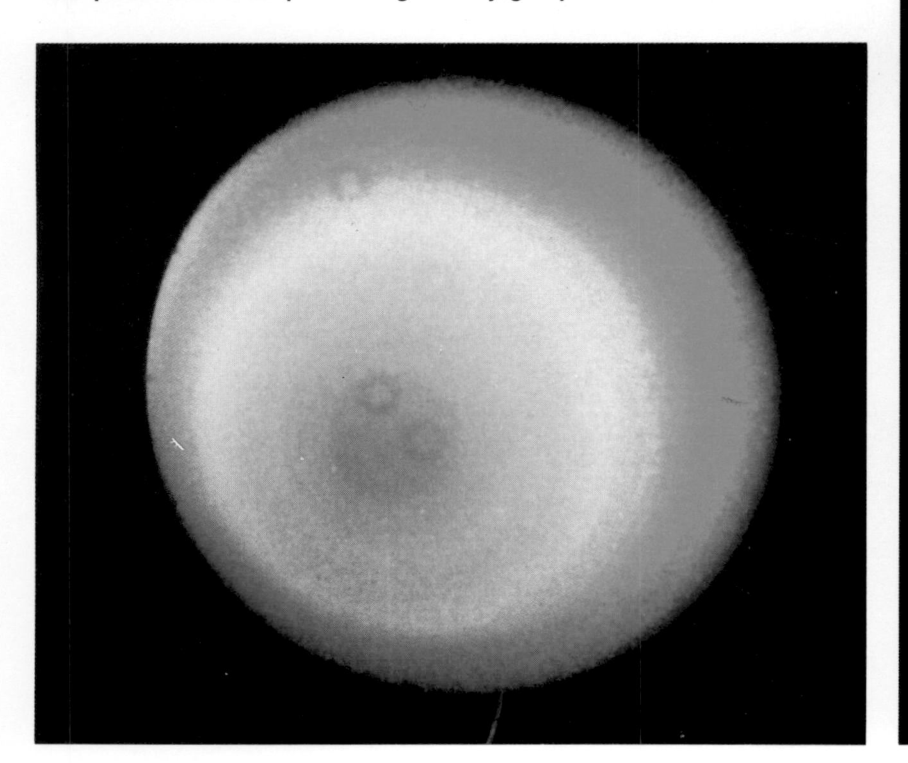

about atmospheric properties and the absence of an intrinsic magnetic field observed by previous spacecraft. It discovered an excess of primordial gases, compared with Mars and earth, that seem to conflict with theories of planetary evolution.

Pioneer 12

On 5 March 1987 NASA's Pioneer 12 spacecraft, orbiting Planet Venus, began six weeks of measurements of the newly discovered Comet Wilson, as the comet streaks by Venus and arcs past the sun.

Controllers at NASA's Ames Research Center, Mountain View, California, will use Pioneer 12 to track Wilson from 14–21 March, then stop for 10 days as the comet dips far beneath the plane of the solar system and Pioneer's field of view.

Pioneer 12 will measure the rate of water evaporation from the comet's nucleus. It also will study the amount of carbon and oxygen emitted, by observing the comet in the ultraviolet portion of the electromagnetic spectrum.

NASA's Deep Space Network

The worldwide NASA Deep Space Network (DSN) provides the earth-based radio communications link for NASA's unmanned interplanetary spacecraft. The DSN has provided telecommunications and data acquisition support for deep space exploration projects since 1961.

Since 1961 the network has provided the vital data link support for the Ranger (1961-65), Surveyor (1966-68) and Lunar Orbiter (1966-67) explorations of the moon; the Mariner missions to Venus (1962 and 1967); Mars (1964, 1969 and 1971) and Venus-Mercury (1973); the Pioneer inward and outward heliocentric orbiters (1965-68) and the Pioneer missions to Jupiter and Saturn (1972 and 1973).

DSN also supported the Pioneer-Venus orbiter and multiprobe (1978); the joint US-West German Helios Sun orbiters (1974 and 1976), the Viking orbiter-lander missions to Mars (1975), the Voyager missions to Jupiter, Saturn, Uranus and Neptune (1977) and secondary support for the manned Apollo lunar landing missions (above dates are launch dates).

The return of scientific data from planetary encounters has been dramatically increased by continuing research and development. For example, in 1965, the data transmission rate for the Mariner 4 spacecraft during its Mars flyby was 8⅓ bits per second. The best television performance was 22 coarsely defined pictures, showing a narrow strip of the planet. Distance between earth and Mars was 400 million km (435 million miles).

The communications record, however, presently belongs to Pioneer 10. Launched in 1971 on a mission to Jupiter, it encountered the planet in 1973 at a range of 827 million km (500 million mi.). Pioneer 10 was then placed on a trajectory that made it the first earth-made object to leave our known solar system. On 13 June 1983, Pioneer was still sending data back as it passed Neptune (Pluto's eccentric orbit now is inside Neptune's) a distance of more than 4.5 billion km (2.8 billion miles).

Although the DSN's primary activity is telecommunications support for unmanned space exploration, the stations are also used as scientific radio telescopes for radio astronomy experiments such as the study of natural radio sources (pulsars and quasars), radar studies of planetary surfaces and Saturn's rings, celestial mechanics experiments including tests of the theory of relativity and NASA's Search for Extraterrestrial Intelligence (SETI).

***At bottom, below:* Illustrations of the Pioneer Venus Multiprobe and Orbiter. The Pioneer Venus Orbiter took this ultraviolet photo of Venus' cloudtops *(below). At right:* Technicians test the Pioneer Venus Orbiter; the Multiprobe is at rear.**

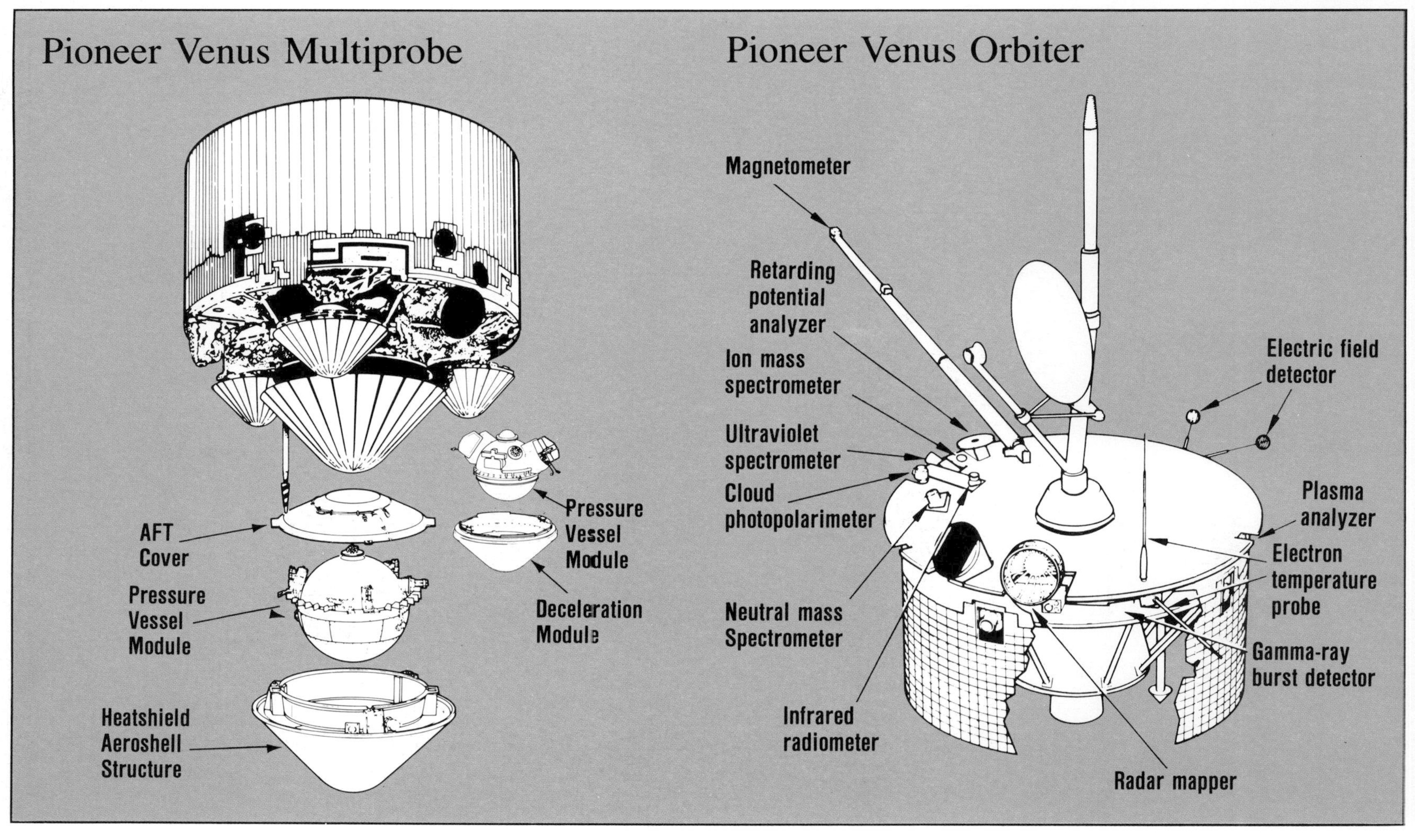

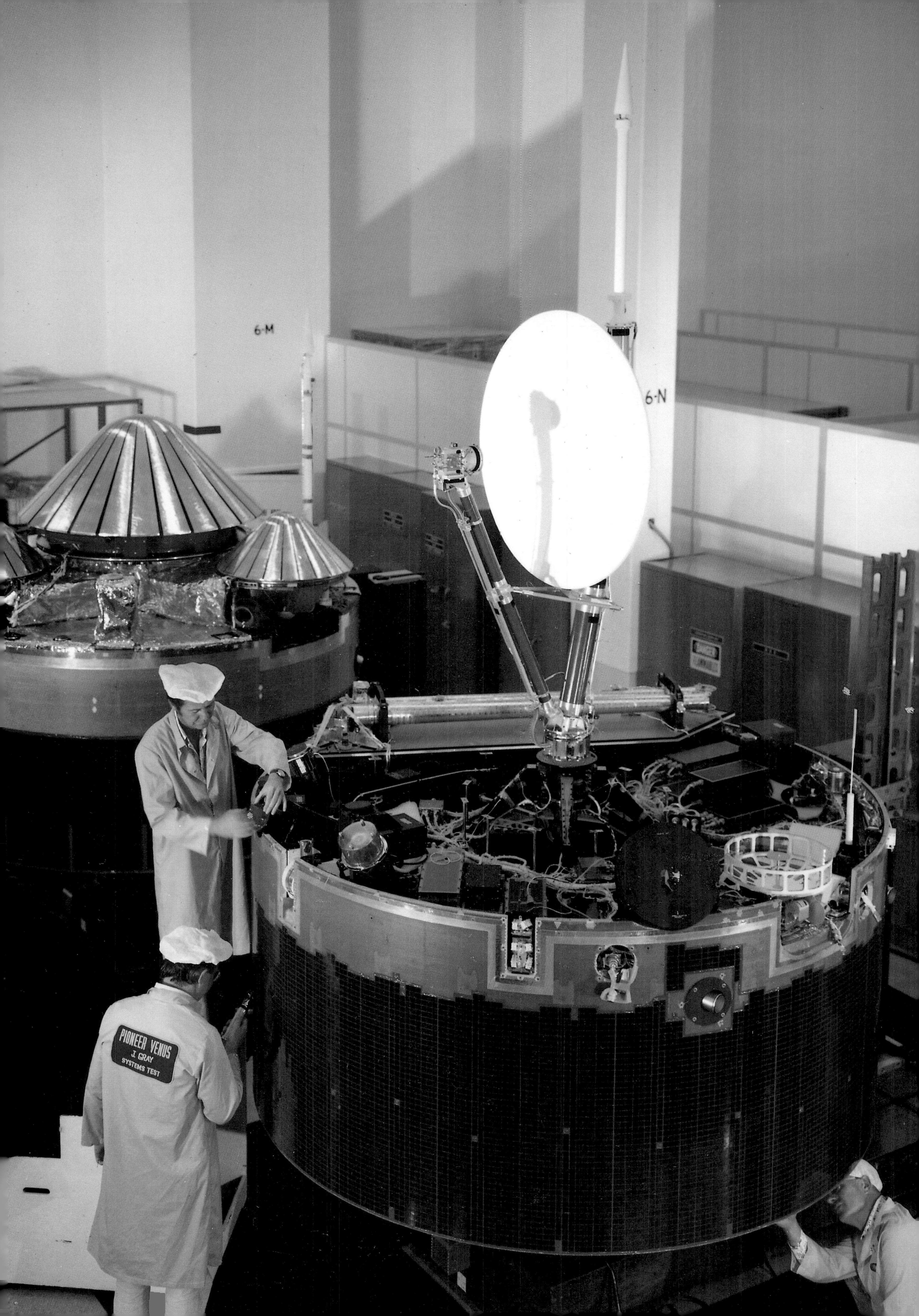
6-M
6-N
PIONEER VENUS
J. GRAY
SYSTEMS TEST

SPACE APPLICATIONS

by Victor Seigel

NASA's space applications program, in which satellite technology is directly applied to benefit people, has been a driving force for human progress. Its value is incalculable and steadily growing.

The Landsat Program

For example, individuals and organizations of more than 100 nations are employing pictures and other data from Landsat Earth Resources Satellites in a diverse number of areas. Among these are agriculture, forestry, land use management, cartography, geology, hydrology, hydroelectric siting and irrigation planning, environmental protection, flood-damage assessment, prospecting for minerals and hydrocarbons, coastal zone management, urban planning, beach-erosion forecasting, siting of offshore facilities and snow-mass mapping and spring run off prediction. With Landsat, the whole earth can be rapidly and repeatedly surveyed at minimal cost. Landsat helps us to discover, inventory, and manage our renewable and nonrenewable resources, alerts us to environmental dangers and keeps us abreast of natural and manmade changes on earth's surface.

A remarkable achievement of our space program is the capability it has provided for keeping a current inventory of our resources, monitoring the quality of our environment and maintaining up-to-date maps depicting natural and manmade surface features at minimal cost. Steady advances in such a capability have already been achieved through NASA's Landsat program.

The advantages of a satellite system like Landsat over aerial and surface surveys, which continue to supplement satellite observations, are significant. Landsat can make repeated observations of an area, revealing changes. It can acquire data from areas where it would be too hazardous or inordinately expensive to do so by other means. It can also reveal large scale features that are overlooked when viewed at lower altitudes.

NASA launched Landsat 1 (originally Earth Resources Technology Satellite, or ERTS 11) on 13 July 1972. It launched Landsat 2 on 22 January 1975; Landsat 3, 5 March 1978; and Landsat 4, 16 July 1982.

Among the areas to which Landsat contributes are agriculture, cartography, water management, flood damage assessment, environmental monitoring, rangeland management, urban planning and geology.

Landsat rapidly provides conventional and near infrared pictures of urban areas, states or regions for land-use planners; surveys rangelands to monitor availability of forage for livestock and to track the animals; observes water levels in reservoirs and snow cover in mountains to help forecast water availability for hydroelectric, agricultural and home use.

It also reports on faults and other geologic formations that hint at mineral, oil, gas or coal accumulations; quickly inventories many fields of different crops in California's Imperial Valley; and describes offshore currents to help control beach erosion and locate offshore facilities. The above are a few examples of hundreds of tests and demonstrations of Landsat potential.

In November 1979, NOAA was designated as manager of all civilian remote sensing activities and directed to pursue eventual commercialization of the Landsat system. NASA continues research and development aimed at advancing Landsat capabilities. On 31 January 1983, NASA turned Landsat 4, the latest and most advanced of the series, over to NOAA.

Landsat Earth Resources Satellites survey the earth's surface photographically, and are invaluable aids to environmental management. *At right:* The 22 January 1975 launch of Landsat 2, which was equipped to photograph the earth in four spectral bands. *At far right:* Landsat 3, as it underwent tests previous to its 5 March 1978 launch from the Western Test Range. This was the first Landsat that was equipped to photograph earth resources in total darkness.

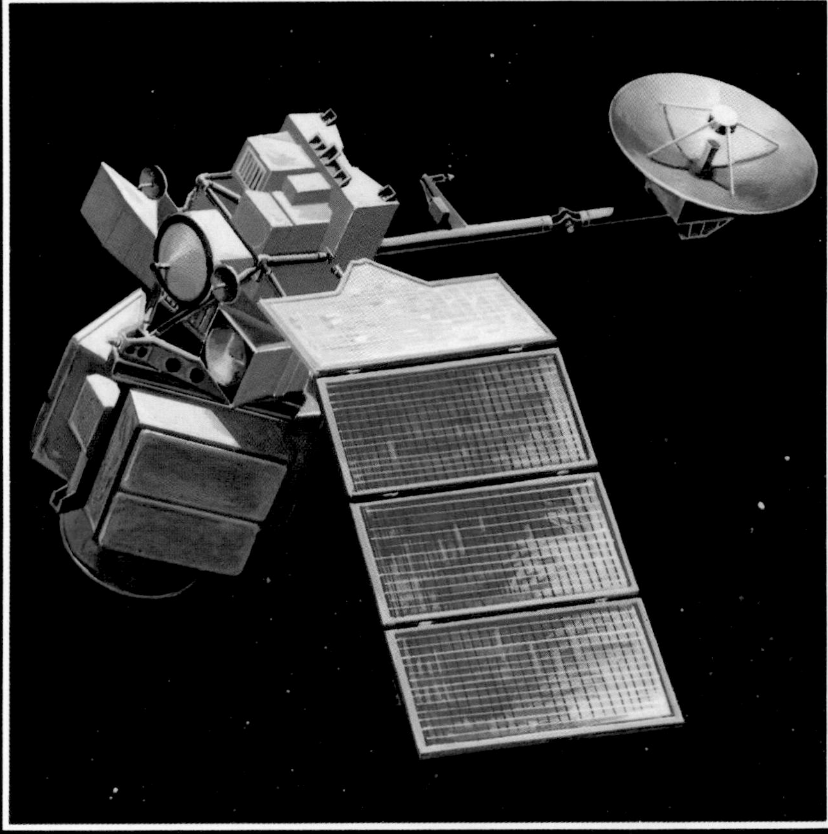

The photo *above* is an artist's concept of Landsat 4 in orbit. *Below:* A Landsat 4 false-color photo of New York City. False color imaging is used to enhance certain kinds of detail, as in the Landsat 4 composite photo of the San Francisco Bay Area *at below right. At right:* A Landsat 4 infrared photo of Death Valley, in California and Nevada. *At left* is a mockup of a Landsat satellite above the earth.

Communication by Satellite

Satellite communications give people front-row seats to historic events regardless of where they occur. They have reduced long distance costs and made available a host of new communications services. They have left an indelible mark on the world and forever changed world information patterns.

In 1960, millions throughout the world began to view a bright star moving across the night sky. This moving star not only dramatized America's space program but also was the harbinger of the communications revolution that would change the world.

The star that started it all was Echo 1 which NASA launched 12 August 1960. Echo 1 was a 100-foot sphere made of aluminum-coated polyester film. Radio signals sent from one ground station were reflected off of its shiny surface to another ground station.

The success of Echo 1 led to deployment of Echo 2, a 135-foot-diameter sphere with a tougher (more rigid) skin than Echo 1, on 25 January 1964. Many communications experiments were carried out with Echos 1 and 2. In addition, the satellites were used as reference points in geodetic observations. The technology of inflatable spheres established in Echo was also applied to Pageos, a geodetic satellite, and to a number of atmosphere Explorer satellites.

Echo 1 was followed by a series of rapid developments in communications satellites. On 10 July 1962, NASA launched the solar-powered Telstar 1, built by the American Telephone & Telegraph Co. Telstar was an active-repeater satellite, meaning it received, amplified and retransmitted radio signals, unlike Echo which was a radio signal mirror. This first non-governmental satellite conducted the first live intercontinental television demonstrations between the United States and Europe.

NASA launched its active-repeater Relays 1 and 2 on 13 December 1962 and 21 January 1964, respectively. It conducted thousands of tests and demonstrations of transoceanic and intercontinental public telecasts via these satellites, including the first between Japan and the United States.

In the midst of these rapid advances, the Communications Satellite Act of 1962 was signed on 31 August 1962. The act called for creation of a private commercial communications satellite corporation and, in conjunction and cooperation with other nations, of a global commercial communications satellite system.

Another kind of orbit is called geostationary or geosynchronous. In such an orbit, a satellite circles over earth's equator at an altitude of about 22,300 miles. Because the satellite's orbital period is the same as the period of the earth's rotation (about 24 hours), the satellite remains stationary relative to a point on earth's surface. Three geostationary satellites spaced equidistant around earth could provide global coverage.

The big questions were whether such orbits could be achieved and maintained and distance that radio signals would have to travel would cause a time lag or echo in communication via satellite. NASA technology solved both problems in its Syncom (for Synchronous Communications) satellite experiments.

After an initial failure with Syncom 1 on 14 February 1963, NASA launched Syncom 2 into synchronous orbit on 26 July 1963. Syncom 2 was placed in an orbit with an inclination of 33 degrees. As a result of this inclination, its ground track over earth resembled an elongated figure 8 stretching from 33 degrees north to 33 degrees south latitude with the crossover point on the equator.

Experiments with Syncom 2 demonstrated the practicability of a synchronous orbit. On 19 August 1964, NASA launched Syncom 3 into a

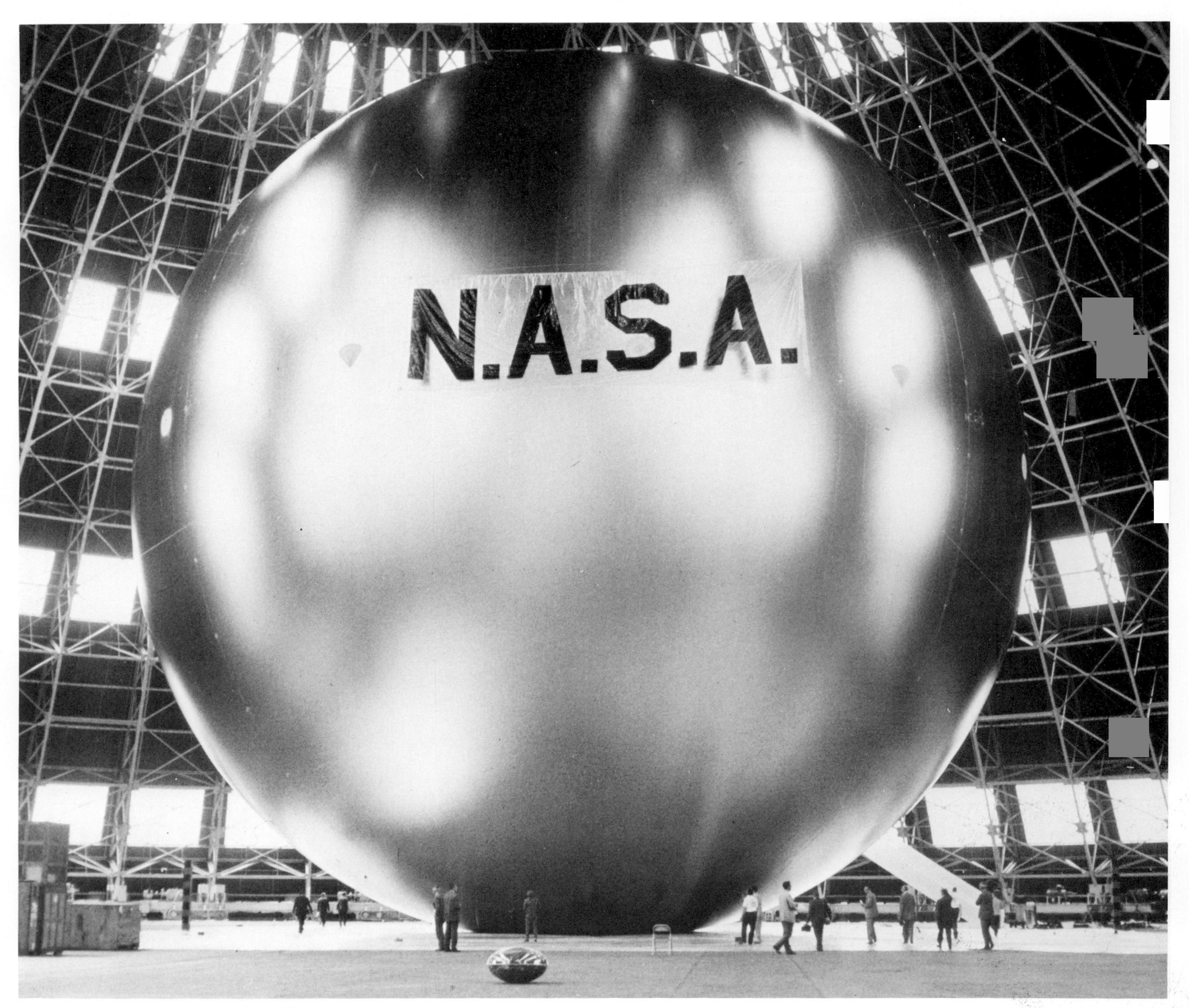

At left: The passive communications satellite, Echo 1, which was launched in 1960. *Below:* Bell Labs/AT&T's Telstar, the first satellite developed and built by a private company, was also the world's first commercial spacecraft. *Above:* A Relay satellite; Relay 1 and 2 served television and voice communications stations in the US, Europe, Japan (with Relay 2) and Brazil. *At right:* A Relay launch.

geostationary orbit. Syncom 3 was the world's first geostationary satellite. Work with Syncom 3 provided much of the additional scientific and engineering data required to lay a basis for the global commercial communications satellite system.

NASA's Syncom and other communications satellites laid the groundwork for INTELSAT, the global commercial communications satellite system in which more than a hundred nations participate, and scores of domestic systems.

NASA technology and demonstrations have contributed to a vast increase in commercial communications service. The first was global TV, which gave people front-row seats to historic events. Among others are the linking of airliners on transoceanic flights with ground terminals and direct broadcast of satellites to rooftop antennas rather than through huge ground stations.

NASA has also shown the potential of communications satellite systems for improving education and health care for persons in remote and isolated areas. By greatly multiplying the number of long distance communications channels available, communications satellites literally revolutionized communications, creating a host of benefits.

Applications Technology Satellites

The versatile Applications Technology Satellite (ATS) program pioneered new technology for weather and communications satellite systems and demonstrated new communications services. The latter displayed the potential of satellite systems for improving health care and education for persons living in remote and isolated areas. ATS were geostationary spacecraft.

ATS 1, launched 7 December 1966, provided the first continuous observation of weather from geostationary orbit. ATS 3, launched 5 November 1967, provided the first color telecasts of earth from geostationary orbit. Together the satellites demonstrated the value of weather satellites in geostationary orbits. This led eventually to NASA's Synchronous Meteorological Satellite program, the prototype of NOAA's Geostationary Operational Environmental Satellite System (GOES).

ATS 1 and 3 far exceeded their design lifetimes and lasted into the 1980s. They have been used for a variety of public services. ATS 1 has served since 1971 to link public health aides in remote Alaskan villages with Public Health Service physicians in Fairbanks and Anchorage, improving health care for native Alaskans. At the same time, it was used for PEACESAT — Pan-Pacific Education and Communications Experiment by Satellite. Initiated by the University of Hawaii, PEACESAT involved 12 Pacific island nations. During a major flood in Alaska in 1967, ATS 1 linked the flood area to US government agencies. During the eruption of Mount St Helens in 1980, ATS 3 relayed messages from an Air Force jeep at the disaster site.

In 1982, the General Electric Co. developed a remote communications terminal small enough to fit into two suitcases and to be carried aboard an airplane. The terminal can be powered from an ordinary AC outlet or car battery and can communicate with any of several ground stations via ATS 3. ATS 6, launched 30 May 1974, was the last and most advanced of the ATS series. It pioneered telecasts via satellite directly to hundreds of small low cost ground receivers.

The key feature of ATS 6 was its 9-meter (30-foot) diameter mesh antenna that could be pointed to within a tenth degree of arc. The antenna enabled ATS 6 to relay strong television signals to suitably augmented small ground receivers. Typically, huge expensive ground stations had been required to receive and amplify faint signals from communications satellites before transmission to home TV sets.

Weather Satellites

Looking at weather from satellites has revolutionized the science of meteorology and improved accuracy of weather forecasts. Weather satellites have contributed to saving countless lives and millions of dollars of property in the United States annually by keeping a constant watch for and on typhoons in the Pacific and hurricanes in the Caribbean and Atlantic.

***At top, right:* An artist's conception of Syncom 2, the world's first successful synchronous orbit communications satellite, which was launched on 26 July 1963. *At right:* A technician makes an adjustment on a Hughes Corporation Telstar 3 spacecraft. *At far right:* Applications Technology Satellite ATS 1 pioneered new communications and weather technology systems, and was launched on 7 December 1966.**

Satellite transmissions have been incorporated into operational weather data since the launch of NASA's experimental Tiros 1 weather satellite on 1 April 1960. Tiros stands for Television and Infrared Observation Satellite. Tiros 1 could provide only daytime pictures of mid-latitude areas of earth. Tiros 2, launched 23 November 1960, was equipped with infrared sensors that enabled it to take nighttime cloud-cover photographs.

NASA improved Tiros over the years. One of the more significant improvements was the inauguration of the Automatic Picture Transmission (APT) system with Tiros 8, launched 21 December 1963. The system enabled weather forecasters everywhere to receive cloud-cover and other data for local area coverage from Tiros with comparatively inexpensive ground equipment.

Tiros 10, launched 2 July 1965, was the first spacecraft funded by the US Weather Service of the Department of Commerce. When the Weather Service became part of the department's Environmental Science Services Administration, Tiros Operational Satellites were designated ESSA. When ESSA was in turn absorbed by the department's National Oceanic and Atmospheric Administration, Tiros satellites were designated NOAA. Under a NASA-Department of Commerce agreement, NOAA establishes requirements for and operates weather satellites and NASA develops the spacecraft and associated ground stations, launches the satellites and checks them out before turning it over to NOAA.

Today's Tiros N (NOAA) satellite system is a far cry from Tiros 1. It covers the globe in a low polar orbit about four times daily. It provides 24-hour coverage of cloud-cover, earth surface and cloud-top temperatures, vertical temperatures and moisture profiles of the atmosphere, meteorological data from fixed and floating platforms (balloons, buoys and unattended stations in remote areas), and data on changes in the Van Allen Radiation Region. Such changes are related to changes in solar activity and contribute to transfer of solar energy between the sun and earth.

NASA's Synchronous Meteorological Satellites (SMS) 1 and 2, launched 17 May 1974 and 6 February 1975, respectively, were immediately used by NOAA and were the prototypes of its Geostationary Operational Environmental Satellite (GOES) system. As with NOAA satellites, NASA develops, launches and checks out GOES before turning it over to NOAA. At least two working GOES are maintained in orbit, assuring constant watches on the Atlantic and Pacific approaches to the Western Hemisphere and on hemisphere land masses to warn of hurricanes and other destructive storms.

They also locate Gulf Stream and other currents for shipping and fishing, warn Florida citrus growers of the approach of crop-killing frosts and provide a large variety of information such as atmospheric profiles of temperature and moisture, sea surface temperature, solar activity and magnetic field data that is crucial to accurate weather forecasting.

GOES 1 was launched 16 October 1975; GOES 2, 16 June 1977; GOES 3, 16 June 1978; GOES 4, 9 September 1980; GOES 5, 22 May 1981; and GOES 6, 28 April 1983.

NASA's experimental Nimbus satellite program contributed many new instruments and techniques to weather and environmental satellites. Each

Below: **Tiros 8, shown here in a mockup, was the first satellite with Automatic Picture Transmission technology.** ***At right:*** **Tiros Operational Satellite TOS E, aka ESSA 7 when it achieved orbit.** ***Below opposite:*** **A Tiros N satellite.** ***At below right:*** **NOAA 7.** ***Above opposite:*** **Synchronous Meteorological Satellite SMS 1.**

Nimbus introduced new technology or new techniques while improving on existing technology. As examples, Nimbus 1, launched 28 August 1964, tested advanced optical and infrared cameras; Nimbus 2, launched 15 May 1966, demonstrated that infrared pictures from weather satellites could be read out live with automatic picture transmission (APT) equipment. Nimbus 3, launched 14 April 1969, tested equipment to make vertical measurements of temperature and moisture, to describe ozone distribution, to gather solar radiation data, and acquire data from sensors on fixed platforms and moving platforms. Nimbus 4, launched 8 April 1970, studied ozone deterioration and reported that between 1970 and 1972, it was only about a half that predicted.

Nimbus 5, launched 11 December 1972, improved techniques to forecast tropical storm movement. Nimbus 6, launched 12 June 1975, demonstrated gathering information from moving platforms. Its biggest effort involved relaying data from about 1000 fixed, floating and airborne sensors to investigators in Australia, Brazil, Canada, France, Norway, the Republic of South Africa and the United States.

Nimbus 7, launched 24 October 1978, showed how satellites could help commercial airliners avoid heavy ozone concentrations which can adversely affect passengers. During the 1970s, Nimbus participated with Tiros in the Global Atmospheric Research Project. This was a joint project of the World Meteorological Organization and the International Council of Scientific Unions.

Nimbus contributed to such operational activities as providing sea ice data for the US Navy and relaying to oilmen icepack movement data from sensors airdropped on the Beaufort Sea icepack north of Alaska's Prudhoe Bay. Nimbus data was also used to prepare global rainfall and ozone distribution atlases.

Analyses of Nimbus data led to many discoveries. One was related to the amount of solar radiant energy reaching earth. Scientists assumed this amount remained more or less level and gave the phenomenon the name of solar constant. Nimbus 7 data indicated the solar constant was not constant. There were not only short term variabilities of between one-tenth and two-tenths of a percent but also an overall downward trend of two-hundredths to four-hundredths of a percent per year in the amount of solar radiant energy reaching earth.

Information from weather satellites has been incorporated into operational weather data since launch of NASA's Tiros 1, the world's first meteorological satellite. NASA continues to develop ground stations and develop and launch weather satellites as a partner with NOAA which establishes requirements for and operates the satellite system.

NASA has expanded weather satellite system capabilities from the original cloud pictures to acquisition of vertical temperature and moisture profiles of the atmosphere, sea and land — surface temperatures, cloud-top temperatures, rainfall and moisture patterns, sea ice data, solar radiation, heat balance and other information related to weather and climate. With such information, scientists may be able to develop a numerical model for weather to contribute both to long term (3- to 14-day) forecasts and climatology. This benefit would be added to the many that weather satellites have already given the world — such as increased accuracy of day-to-day forecasts; alerting coastal populations to approaching hurricanes, thus saving countless lives and millions of dollars of property annually; and informing shippers about the locations of the shifting Gulf Stream, enabling them to ride it or avoid it and thus save fuel.

In addition, the NASA-developed Automatic Picture Transmission (APT) system, a relatively low cost way of acquiring pictures and other data from NASA-NOAA weather satellites, is being employed in about 135 countries.

Nimbus 6 *(below left)*, launched on 12 June 1975, performed an important information gathering experiment. *Below:* Geostationary Operational Environmental Satellite GOES 3. *At right:* A satellite typical of the GOES D through F design.

Geodesic Surveys from Space and Other Programs

NASA is employing advanced space technology to make precise measurements of movements of tectonic plates into which the rigid outer shell of the earth is divided. Such movements lead in time to earthquakes and volcanic eruptions along the edges of the plates. NASA is observing plate movements along the plate boundary known as the San Andreas Fault in California and, in cooperative programs with other nations, other earthquake prone areas. The measurements may contribute to an earthquake-forecasting system that could save countless lives annually.

NASA's National Geodetic Satellite Program, now completed, has resulted in improved maps and knowledge of earth's structure. It has contributed to surface and air navigation and to the launch, guidance and pinpoint landing of spacecraft such as the Space Shuttle.

Geodesy involves development of a global network of triangles for accurate determinations of latitude and longitude of any point on earth and distances between points. It also refers to mathematical determinations of the size, shape, and distribution of earth, including variations in gravity. NASA's Explorers 27 and 29 (GEOS 1 and 2, or Geodetic Satellites 1 and 2), launched 29 April and 8 November 1965, and Pageos (Passive Geodetic Satellite), launched 24 June 1966, were devoted to geodesy. Geodetic studies were also accomplished with the ionosphere Explorer 22, Echo 1 and 2 communications satellites, and Army's SECOR satellite.

Information from these programs proved that even the best previous maps frequently showed points many kilometers from where they actually are. The geodetic project, which involved more than 30 nations and included NASA, the Department of Commerce, and the Department of Defense in the United States, has benefited surface and air navigation and contributed to the pinpoint launches, guidance, and landings of the Space Shuttle.

GEOS (Geodynamic Experimental Ocean Satellite) 3, launched 9 April 1975, provided more geodetic and geophysical data about the oceans than were accumulated in all previous years of ship measurements.

The Department of Defense estimated that the data saved them about $140 million in ship survey operations over 10 years. GEOS 3 also increased accuracy of measurements of the ocean geoid—the level that the ocean surface would assume in the absence of wind, currents, tides and gravity anomalies.

GEOS 3 was the first satellite to make precise measurements of the topography of the ocean surface and of sea state—wave height, period and direction. Its information contributed to models for forecasting ocean conditions.

In a technological breakthrough, GEOS 3 demonstrated in 1978 that its radar altimeter for sea measurements could be as accurate in large-scale land contouring as ground and aircraft surveys are for small-scale mapping.

Data from GEOS 3 and other studies were used in the development of Seasat, launched 26 June 1978. It produced a wealth of ocean data before its life was cut short, 105 days after launch, by a power failure. Among them were detailed bathymetric (water-depth) and geologic information for vast areas of the ocean. Its information resulted in many corrections of bathymetric charts.

Lageos (Laser Geodynamics Satellite), launched 4 May 1976, is being used in a 50-year program to help measure the minute movements of large tectonic plates into which the earth's outer shell is divided. Such movements are related to earthquakes, volcanic eruptions, accumulations of oil, gas, coal and minerals and building and break-up over millions of years of mountain ranges and continents.

An early finding of Lageos experimenters is that movements of the earth's surface along California's San Andreas fault are as much as 50 percent faster than the historical average. The faster movements may indicate that strain is building up more rapidly than expected.

The feasibility of mapping variations in surface heat by satellite and the benefits that can be derived from such data were demonstrated by an Applications Explorer satellite in a Heat Capacity Mapping Mission (HCMM). The HCMM Explorer was launched 26 April 1978.

The information from HCMM appears to contribute to many fields including agriculture, hydrology, oceanography, geology, meteorology, environmental monitoring and prospecting. HCMM results are expected to contribute to advances in Landsat capabilities.

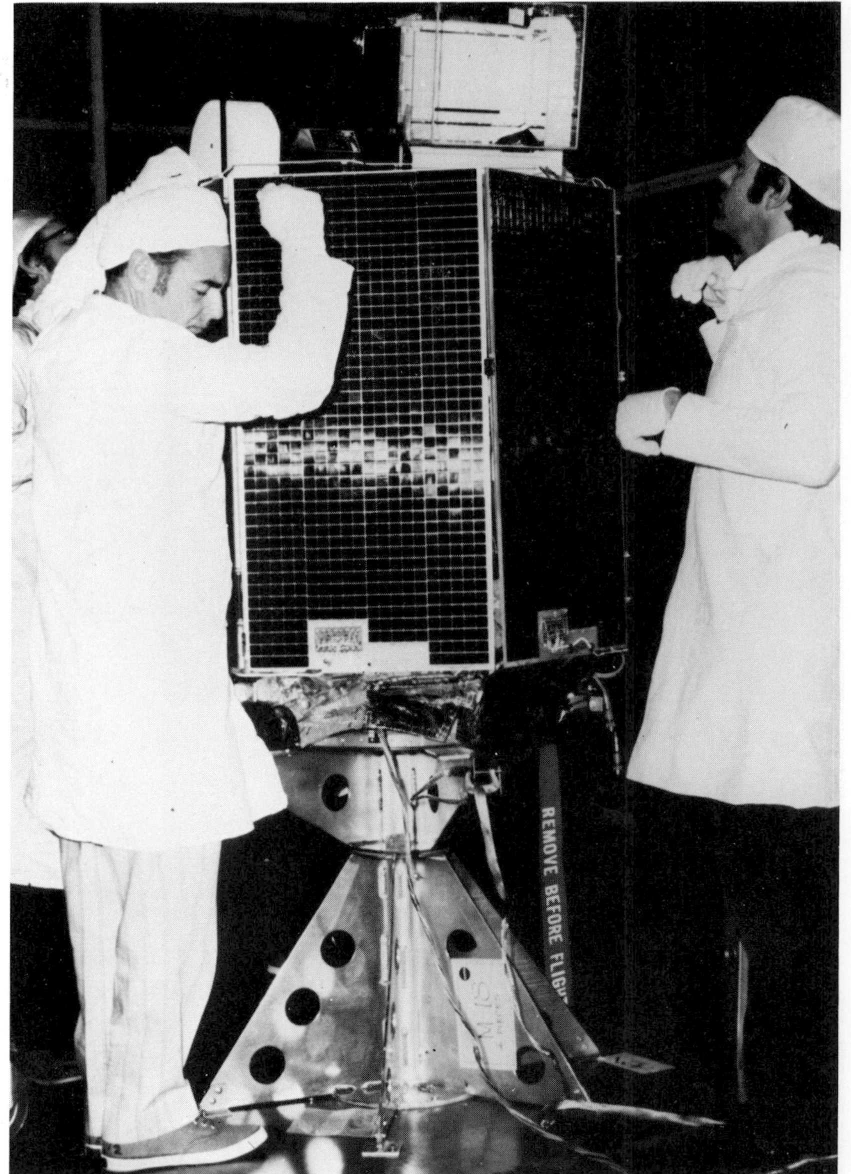

At top right: **Geodynamic Experimental Ocean Satellite 3 was launched on 9 April 1975.** ***At right:*** **The Heat Capacity Mapping Mission HCMM Applications Explorer satellite. The Laser Geodynamics Satellite *(at far right)* was launched on 4 May 1976.**

AERONAUTICS

by Stuart Rosenbaum

When NASA was created in October 1958, it was formed around a nucleus of NACA, the National Advisory Committee for Aeronautics. NACA had been created in 1915 '. . . to direct the scientific study of the problems of flight with a view to their practical solution and to determine the problems which should be experimentally attacked; to discuss their solution and their application to practical questions . . . and to direct and conduct research and experiment in aeronautics in such laboratory or laboratories.'

The National Aeronautics and Space Administration was chartered officially to conduct aeronautical research among its other defined tasks. The National Aeronautics and Space Act of 27 July 1958, the legislation that established NASA, states that the general welfare and security of the United States require adequate provision to be made for aeronautical activities, and that these activities should be so conducted as to contribute materially to one or more of these objectives:

- The expansion of knowledge of phenomena in the atmosphere.
- The improvement of the usefulness, performance, speed, safety and efficiency of aeronautical vehicles.
- The preservation of the role of the United States as a leader in aeronautical science and technology.
- The most effective utilization of the scientific and engineering resources of the United States in order to avoid unnecessary duplication of effort, facilities and equipment.

NASA research scientists, part of a team that includes industry, universities, other government organizations and laboratories, both private and government, work toward those broad objectives. Specifically, NASA aims its research toward the advancement of both civil and military aeronautics, pointing toward new concepts of flight, seeking new approaches to solve the ever evolving new ideas to stimulate the designers of tomorrow's aircraft and aeronautical vehicles.

NASA's broad field of aeronautical research has as its primary subjects the vehicles and powerplants that use the earth's atmosphere for flight. It also is concerned with the aeronautical aspects of space vehicles that depart from or land on the earth.

The major share of this work is done at four NASA centers: Langley Research Center, Hampton, Virginia; Ames Research Center, Mountain View, California; Dryden Flight Research Facility, Edwards AFB, California; and the Lewis Research Center, Cleveland. Additional studies are conducted at other NASA installations or at the laboratories and facilities of other government agencies. Private industry makes important contributions, from self-supported research and development programs and from NASA-funded programs.

NASA aeronautical research is categorized in a quartet of headings: proof of concept, extension of the art, future needs and problem solving.

This is an approach that often, but not always, requires the building and testing of a special aeronautical vehicle. The best known examples are the famed X series of research aircraft developed by NACA and the military services in the 1940s and subsequently. A more recent example is the joint Navy/Army NASA-Bell XV-15, a unique rotorcraft with a potential for both military and civil use. This program exemplifies the means by which the feasibility of any new concept must be embodied and demonstrated in a tangible flying machine before its actual potential can be established.

Extension of the art takes as a basis the contemporary state of that art and builds on that foundation. Today's subsonic transport aircraft are, for example, well understood. But continuing research in aerodynamics, propulsion, structures, materials and avionics indicates that tomorrow's

NASA uses Lockheed F-104s like that *at right* for high altitude research. *At above right,* Space Shuttle *Challenger* touches down at Edwards AFB, as a T-38 chase plane oversees the landing. NASA researches both civil and military aeronautics.

NASA
923
Challenger
NASA
United States
RESCUE
NASA

transports could be vastly improved by the incorporation of new ideas. The art has been extended, and that body of knowledge is available to industry as foundation stones for the next-built generation.

Future needs call for the broadest research goals. While sometimes the end point of such work would seem to have little practical current application, other research is quite focused. An example: The investigation of high temperature resistant structural ceramic materials for use in future gas turbines. Today's engines are highly dependent on expensive, strategic metals not occurring naturally in the United States. Not only would successful ceramic materials reduce our dependency on crucial imported metals, but they could also permit useful higher temperatures to be achieved, increasing engine efficiency and fuel economy. Much research is required to establish the feasibility of such applications and to assist in developing useful materials and manufacturing techniques.

Supersonic Cruise Aircraft Research (SCAR)

NASA researchers worked throughout the 1960s on technologies for supersonic transport. By 1971, Boeing's Supersonic Commercial Air Transport (SCAT) was ready for production, but concerns about noise, economy and pollution prevented further funding.

Convinced that supersonic transport research could eventually pay off, in 1973 the government funded the Supersonic Cruise Aircraft Research (SCAR) program. Nine years of a sustained, focused technology program involving NASA and major American propulsion and airframe companies resulted in significant improvements over earlier supersonic transport concepts. By the early 1980s, the SCAR program had developed technologies permitting a greatly increased range, greater passenger capacity, lighter weight and cleaner, quieter, more efficient engines than any existing supersonic transport aircraft possessed.

By 1971, Boeing was ready to produce its Supersonic Commercial Air Transport *(at right)*, but noise, economy and pollution concerns stopped project funding. Taking that into consideration, NASA's Supersonic Cruise Aircraft Research generated such as the advanced, variable-wing supersonic airliner conception shown *below*.

BOEING

The X-15 Program

This NASA program began on 25 March 1960 and terminated on 24 October 1967. The X-15, a 50-foot long, black, stub-winged, rocket-powered flight research craft with a conventional nose-wheel and retractable skids mounted at the rear for landing, was a true aerospace vehicle. With wings and aerodynamic controls it traveled like an airplane in the atmosphere, but in flight beyond the atmosphere, like a spacecraft. In addition to the rocket motor for propulsion housed in the craft, reaction control rockets were also provided to control the craft at extreme altitudes where there was no atmosphere to provide aerodynamic forces on the various control surfaces of the X-15 to effect a change in course or attitude.

The X-15 was launched from beneath the wing of a B-52 at an altitude of 45,000 feet. After its drop, the main rocket engine was fired and the craft climbed in a steep trajectory, then nosed over to descend in an unpowered glide to a landing.

Through a series of progressive steps, the X-15 set new altitude (more than 67 miles) and speed (6.7 times the speed of sound) records. Its 199 flights contributed important data about weightlessness, aerodynamic heating, atmospheric entry, the effect of noise on aircraft materials, and piloting techniques, valuable to the manned space programs which followed the X-15.

The X-15 was a joint NASA/Air Force/Navy project. First piloted by A Scott Crossfield, both Neil A Armstrong, Commander of Apollo 11, and Joe Engle, Commander of STS-2, were among the pilots who flew the X-15 into unexplored areas of flight.

Below: NASA, US Air Force and company test pilots, and one of the three X-15 rocket planes that earned several pilots astronaut wings. _At below far right_ is the X-15A2, with jettisonable external fuel tanks. _At above far right:_ NASA YF-12s, cousins of the US Air Force SR-71 Blackbirds. The far plane with test gear is a YF-12A; the near plane is a YF-12C. These former US Air Force interceptors were given to NASA as hypersonic test beds. _Above right:_ A YF-12 pilot climbs aboard.

06937
NASA
U.S. AIR FORCE
NASA
66671
USAF
U.S. AIR FORCE
APU EXHAUST

These pages: An ER-2 NASA earth resources survey plane. The ER-2's design is related to that of the high-flying TR-1 tactical reconnaissance plane, which itself was based on the U-2 'spy plane' variant, the U-2R. The wing pods of this ER-2 can carry two tons of equipment and sensors.

HiMAT

Highly Maneuverable Aircraft Technology was a NASA/Air Force flight research program to study and test advanced fighter aircraft technologies.

The HiMAT vehicle was a 44 percent scale model with wing-tip-mounted winglets and a small forward canard wing for high maneuverability. It consisted of a core design to which modular components could be attached easily and replaced, a format that allows low-cost testing of a variety of concepts.

In 1979, the remotely controlled research aircraft made its first flight. The following year it achieved near-maximum design maneuverability at sustained near-supersonic speeds, and in 1981 its flight testing was expanded to transonic speeds.

The HiMAT flight test program ended in January 1983. The vehicles had performed superbly, demonstrating twice the maneuverability at transonic speeds of modern fighter aircraft. The large quantities of high quality data obtained in the program will be used in applying new technologies to future aircraft.

Tilt Rotor Research Aircraft

The Tilt Rotor Research Aircraft, (TRRA) XV-15 employs two wing tip mounted rotors to combine the advantages of a helicopter's vertical lift and maneuverability with an airplane's cruising speed. In the air, the rotors tilt forward to become propellers for cruising. This versatile aircraft can take off and land vertically, hover and fly forward, sideways or backwards.

The tilt rotor shows promise in military applications and possibly as a commercial commuter liner operating out of close-to-city heliports. In 1981 the tilt rotor completed the proof-of-concept flight research phase. It flew twice as fast as a helicopter with equal fuel consumption rate and achieved a top speed of 346 mph.

Composite Materials

Unnecessary weight adds to the amount of fuel needed for flight, so the Aircraft Energy Efficiency program has been developing technology for new lightweight composite materials for airframe construction.

Conventional aircraft are constructed primarily with alloys of aluminum, magnesium, titanium and steel. The new composite materials consist of graphite, glass or Kevlar fibers arranged in a non-metallic matrix, generally epoxy. Through correct arrangement of the fiber orientation, the great strength and stiffness of these materials can be applied in arbitrary directions with minimum structured weight. These light, yet strong and stiff materials offer possible weight reductions of 25 percent or more. Beginning with secondary structures not critical to flight safety, some new materials have been flight tested.

The goal is to monitor the materials in daily use on a commercial airline, where the normal wear on the pieces can be observed. Because they replace metal parts on aircraft in service, each new part will be certified by the Federal Aviation Agency (FAA). Eventual testing of a complete wing and fuselage will provide a design base for future energy efficient aircraft.

Crash Dynamics

Recent studies have included an investigation of airplane crash-dynamics information with the intent of increasing the survivability of passengers in an accident. For several years NASA has been deliberately crashing extensively instrumented aircraft, both single and twin-engined, under controlled conditions, to determine exactly how structures behave.

The planes, containing anthropomorphic dummies harnessed in the crew and passenger seats, are crashed onto a runway from a test rig. The data collected helps researchers understand how an aircraft absorbs the energy of impact and how the impact shock to the occupants can be reduced. The tests include the study of improved, load-limiting seats, harnesses, and crushable subfloor and fuselage structures.

A joint project of NASA's Ames Research Center and the US Army's Research and Technology Laboratories, the XV-15 Tilt Rotor Research Aircraft (TRRA) can take off vertically, can fly forward, backward or sideways, and is the most efficient vertical lift aircraft for distances over 100 miles. ***At right:*** **A TRRA on display.** ***At far right:*** **Liftoff!** ***At top right:*** **Hovering near the Jefferson Memorial, and adjusting the tilt motors** ***(at middle right)*** **for level flight** ***(above far right).***

XV-15
TILTROTOR
N702NA
2

Fireworthiness

In a related effort, NASA researchers at Ames and NASA's Johnson Space Center in Houston, are developing fire resistant materials for use inside cabins. One concept uses fire resistant wrappings over conventional polyurethane foam cushions. Another fire resistant, lightweight polymide seat cushion has been developed at Johnson and is being evaluated in service by three airlines. Similar lightweight fireworthy materials are being applied to ceiling, wall and floor panels.

Less flammable jet fuels are also under study, most notably the British developed anti-misting kerosene (AMK) safety fuel, FM-9. A full-scale crash test is planned whose objectives include a test of the new fuel's anti-misting ability to prevent major fires caused by ignition during and after a crash. Along with the Federal Aviation Agency, NASA has been testing the safety fuel and evaluating its compatibility with the most common engines in service.

Automated Pilot Advisory System

For general aviation pilots operating out of small uncontrolled airfields, NASA designed and successfully demonstrated the Automated Pilot Advisory System (APAS) to provide weather, traffic and airport information. The Automated Pilot Advisory System includes a tracking radar, weather sensors, a computer and a transmitter.

A computer-generated voice broadcast of information on air traffic within three miles of the airport is made every 20 seconds, while every two minutes, information is provided on airport identification, active runway, wind speed and direction, barometric pressure and temperature.

Stall Spin Research

The stall/spin phenomenon has been an important cause of accidents in general aviation. A stall occurs when the angle of attack of the wing increases to the point where air across the wing separates instead of following the upper surface; this causes a loss of lift. Following a stall, an airplane sometimes will begin to spin while descending rapidly. Stall/spin research has ranged from early experiments with models in wind tunnels to more recent use of simulators and full-scale flight research vehicles.

In the 1970s, a large-scale effort focused on vertical tail designs and went on to develop a number of wing leading-edge modifications. These modifications have been shown to make certain test airplanes significantly more resistant to and recoverable from spins.

The stall/spin research has produced a large body of data that aids industry in the design of safer airplanes. Continued effort is underway to find ways to increase spin resistance of light aircraft, and to provide analytical techniques for use in design.

Icing Research

An increasing demand for all-weather flights brought on by advances in avionics systems, has brought a renewed interest in improving aircraft safety under icing conditions. Current research is aimed toward developing lightweight, low-power consumption, cost-effective ice protection systems. Analysis, wind tunnel testing and flight research are being used to validate the effectiveness of these protection systems.

Below: **A view of the wind tunnel at Ames Research Center, with retouching to include the test section built in the early 1980s.** ***Below right:*** **A rollout photo of the Grumman X-29 Forward Swept Wing aircraft, built for NASA and the US Air Force and tested at Ames in the mid-1980s.** ***Below, at bottom:*** **The X-29 on its maiden flight of 14 December 1984.** ***At right:*** **The X-29 shows its unique configuration.**

MANNED SPACE FLIGHT

by Victor Seigel (1961-1983) and Bill Yenne (1983-1986)

On American Independence Day, 4 July 1982, NASA astronauts Thomas K Mattingly II and Henry W Hartsfield Jr landed the Space Shuttle orbiter *Columbia* on a concrete airstrip at Edwards Air Force Base, California. Their nearly flawless eight-day earth-orbital mission successfully concluded the series of four orbital flight tests of NASA's Space Transportation System (STS), which was then declared operational. President Reagan, who was at Edwards to greet the returning *Columbia* and its crew, likened the flight to the historic 'golden spike' that inaugurated transcontinental railroading.

Declaring STS operational on Independence Day was symbolic because the system literally frees people to perform economically in space a host of beneficial activities that were once considered impracticable. Perceived in the early- to mid- 1980s as the world's first operating spaceline, STS, was designed to provide regular trips for people and cargo between the ground and earth orbit.

It also serves as a platform in space from which to launch and retrieve spacecraft, conduct experiments, make observations and assemble large structures such as space stations, multi-purpose space platforms, solar-powered electric-generating facilities and huge erectable communications antennas that can lead to another quantum jump in world communications.

The Space Shuttle represents a revolutionary departure from the single-service cone-shaped bodies of Mercury, Gemini and Apollo. The cone-shaped design was chosen because it provided the highest ratio of payload to weight; the blunt ends generated maximum lift and drag during reentry and flight through the atmosphere, and they permitted use of available ballistic missile experience.

The heat shields of these craft were made of a material that was destroyed as it protected the craft from the reentry inferno. In contrast, the Shuttle orbiter is covered by tiles designed both to protect the craft from heating and to last from flight to flight. This sweeping change, involving perfecting the tiles and bonding them to the craft, was among the new technological frontiers that had to be crossed in Shuttle development.

Organized in 1958, Mercury was America's pioneering manned space flight program. Even today, with the accumulated experience of 25 years, engineers and scientists recognize that manned space flights can never be considered routine and that unexpected hazards can occur. At the time of Mercury, however, the known and unknown problems of space appeared to be formidable.

Because space has no appreciable atmosphere, one side of a spacecraft can bake in the sun's heat while the dark side freezes in subzero cold. Tiny micrometeoroids, speeding through space at thousands of kilometers per hour, can pierce a spaceship's hull or an astronaut's pressure suit. Radiation can be lethal to people and degrade materials. The high vacuum of space can kill an exposed astronaut within 30 seconds.

Apparent weightlessness in space can upset delicate biological processes and impair vital organs. It causes liquids to crawl upwards in their containers and escaped liquid to drift about in a cabin, posing hazards to people and equipment.

Yet, in the 20 years between 1961 and 1981, advances in technology and life sciences have, for the most part, enabled America to cope effectively with the problems of manned space flight. America rapidly forged ahead, from the first hesitant steps in Mercury to the nearly normal accommodations afforded by the Space Shuttle.

The Mercury program of one-man spacecraft demonstrated that man can live, eat, work and sleep in space. It also showed that man can augment data acquired from automatic equipment and can make observations from space.

At right: **The 5 May 1961 launch of the Mercury Redstone-3 mission, during which Alan Shepard Jr became the first American astronaut.** ***At far right:*** **STS 6 blast off—the Space Shuttle *Challenger* blasts off from Launch Pad 39A on 19 April 1983.**

USA
NASA
Challenger

In 1961, shortly after the first Mercury suborbital flight, President Kennedy proposed to Congress the bold new initiative that came to be known as Apollo: 'I believe that this nation should commit itself to achieving the goal, before this decade is out, of landing a man on the moon and returning him safely to earth.'

The space flights in the two-man Gemini spacecraft in 1965 and 1968 provided mastery of technology and skills that were crucial to Apollo: maneuvers in space, rendezvous and docking with another vehicle in space, extravehicular activities (EVA), (EVA) and demonstration that man could function effectively in space for as long as two weeks with no lasting harmful aftereffects. In addition, photographs and other data acquired during Gemini's orbital missions provided a wealth of information related to earth's geography, environment and resources and to astronomy.

Even as Gemini was advancing manned space flight technology, a series of manned spacecraft were reconnoitering the moon for Apollo. The Ranger unmanned lunar impact craft provided the world's first lunar close-ups, revealing features not visible to earth observatories, before it crashed as planned on the moon. Lunar Orbiters provided detailed lunar topography that contributed to selection of Apollo landing sites. It also provided tracking, gravity and lunar environmental information. Surveyors, which landed softly on the moon, demonstrated that their landings raised no dust cloud, showed that the moon's bearing strength could support the weight of the Apollo landing craft, provided panoramas of the surrounding moonscape, and showed also that liftoff from the moon would not be periled by a dust cloud. Meanwhile, unmanned Explorer and Pegasus meteoroid-study earth satellites furnished data that contributed to Apollo design.

One of humanity's greatest achievements may well be the comparatively brief period of time in which skills were established; people, materials, and equipment organized; and vehicles and facilities built that resulted at 10:58 pm EDT, 20 July 1969, in the following message from Neil A Armstrong as he stepped from the Apollo lunar module, *Eagle*, and onto the moon's ancient untrodden surface:

'That's one small step for a man; one giant leap for mankind.'

The United States referred to Apollo as a triumph of humanity. For the moment, the triumph gave humanity a new dimension of pride, of togetherness and of confidence that it could achieve other worthwhile but difficult objectives.

Apollo became a scientific mission that vastly expanded knowledge about the moon and earth. Six Apollo expeditions explored the moon, the last in December 1972. Astronauts worked effectively, even drove a vehicle about the lunar surface. They set up experimental equipment that would keep sending data long after they left. They brought back a large variety of lunar samples, countless pictures and other data that would be studied by hundreds of scientists in the United States and abroad for years to come.

Skylab was America's first space station. The Skylab workshop, where three American astronauts lived and worked for long periods, was a modified empty third stage of the Saturn 5 launch vehicle that sent Apollo to the moon. Attached to the stage were an airlock enabling the crew to

Below: Neil Armstrong and David Scott in the Gemini 8 capsule. _At right:_ The 3 March 1969 launch of Apollo 9, the first manned flight of the complete Apollo lunar spacecraft. _Below right:_ Astronaut James Irwin, with the Lunar Roving Vehicle and Apollo Lunar Module _Falcon_ during Apollo 15, in July–August of 1971.

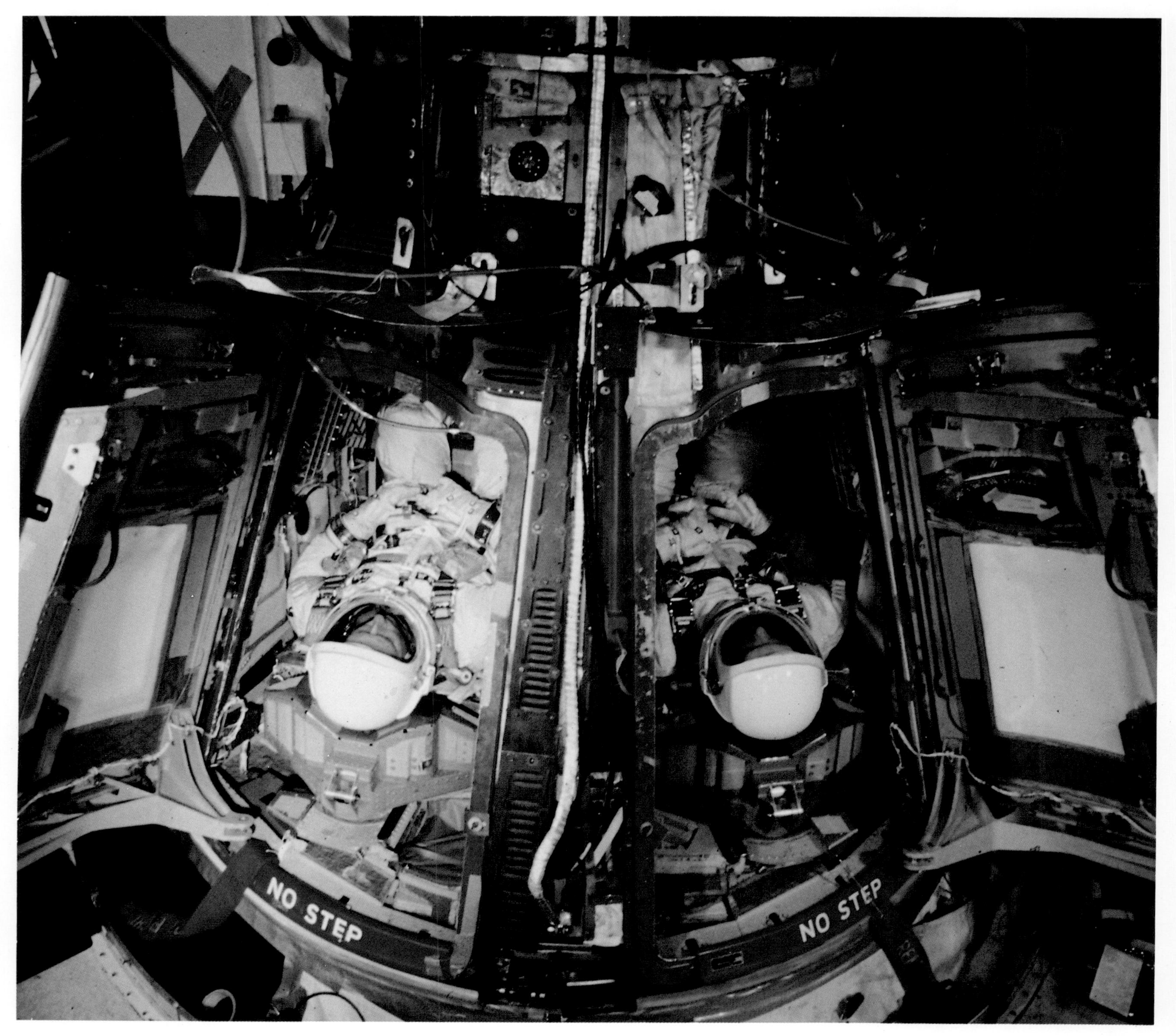

UNITED
STATES

enter and leave the workshop, a multiple docking adapter for parking the Apollo command/service module that was used to travel between Skylab and earth, and the Apollo telescope mount, an astronomical observatory to study the sun. Apollo was launched by a Saturn 1B, the vehicle used in early Apollo earth-orbital tests. Skylab and its first crew were launched in May 1973. Three separate Skylab missions, lasting 28, 59 and 84 days, were conducted until February 1974 when the last crew returned to earth.

Skylab provided a treasure trove of earth survey and solar pictures. It conducted a large variety of tests of industrial and biological processes for manufacturing products in space that would either be unique or of higher quality or produced in greater quantity than possible on earth. Studies of astronaut blood and other samples accumulated and returned to earth advanced knowledge about human biochemical changes in space and, indirectly, human life processes on earth. In Skylab, the astronauts clearly established that with proper equipment, nutrition and exercise, man could work for prolonged periods about as effectively in space as on earth and suffer no lasting harmful aftereffects upon return to earth.

Perhaps the most significant result of Skylab may be the knowledge that people can repair, adjust and install new equipment in space. This ability was quickly demonstrated when the first crew made repairs and installations which saved Skylab from a total loss. Their success supports the plans of Shuttle astronauts to repair satellites and assemble large structures in space.

Skylab was followed in July 1975 by the Apollo Soyuz Test Project (ASTP). ASTP, the world's first international manned space mission, employed the reliable Apollo command/service modules and the Soviet Soyuz.

It tested compatible docking systems as a possible lead to an international space rescue capability and future international manned space missions. It also conducted a large variety of experiments in earth survey, astronomy, life sciences and industrial and pharmaceutical processing.

Apollo-Soyuz ended America's era of expendable manned spacecraft. Six years later, American astronauts returned to space with the revolutionary Space Shuttle, vastly expanding America's capabilities to use space for the benefit of all mankind.

At right: Skylab. _Below:_ Astronaut Thomas Stafford and Cosmonaut Aleksei Leonov in the hatchway that joined the Apollo (near, in the illustration _at far right_) and the Soyuz (also _at far right_) spacecraft for the Apollo-Soyuz Test Project.

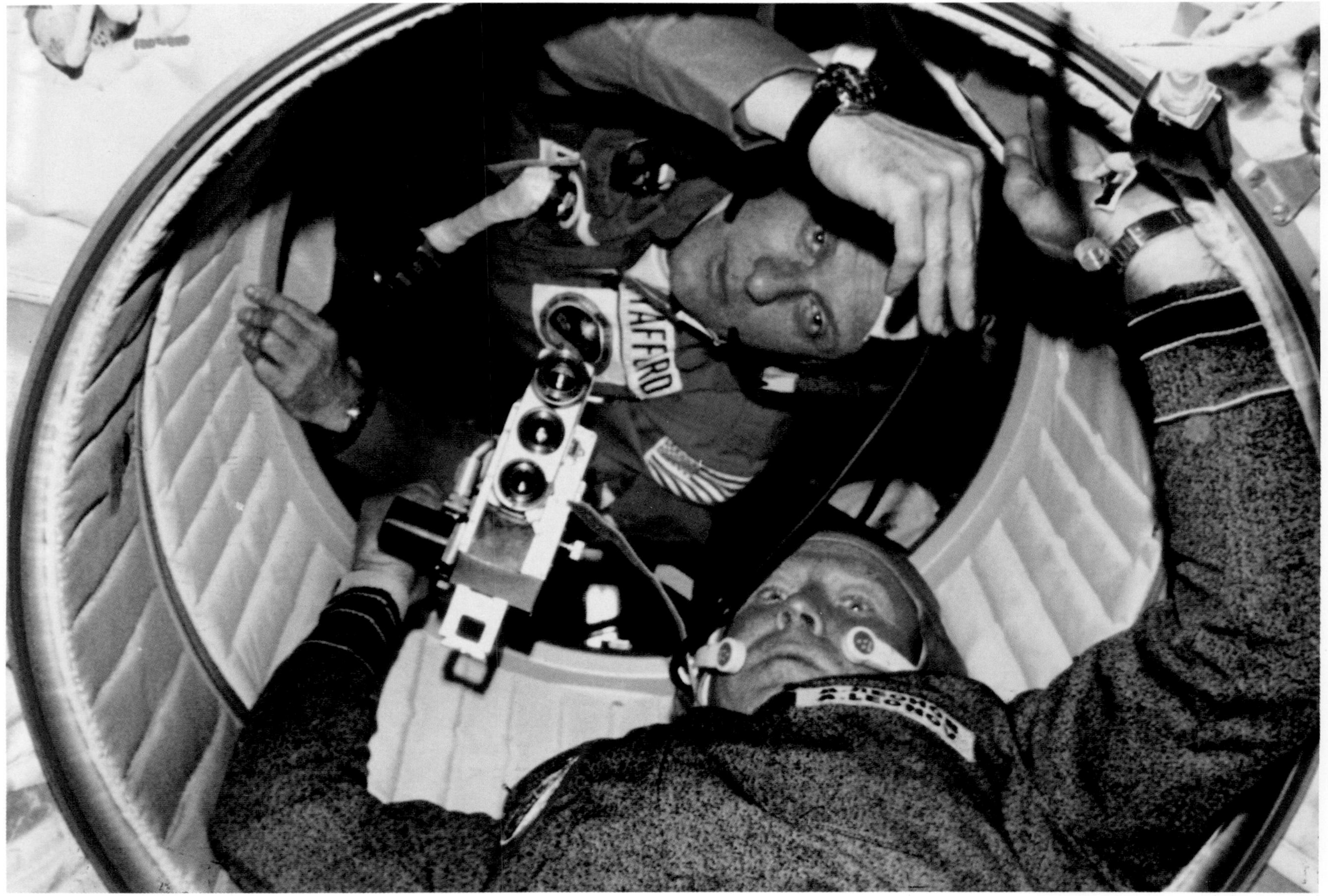

UNITED
STATES

The Mercury Program

Mercury was organized 5 October 1958—just four days after NASA was chartered—to orbit a manned spacecraft, investigate human reaction to and abilities in space, and to return the Mercury spacecraft and its occupant safely to earth.

Late in 1958, the NASA Special Committee on Life Sciences established qualifications for the first Americans to fly into space: a degree or equivalent in physical science or engineering, graduation from a military pilot test school, 1500 hours of flying time including a substantial amount in high performance jets. NASA initially screened the records of 473 military service officers and on 9 April 1959, announced the Mercury astronauts: M Scott Carpenter, L Gordon Cooper, John H Glenn Jr, Virgil I Grissom, Walter M Schirra Jr, Alan B Shepard Jr, and Donald K Slayton.

Little Joe was a series of tests between 1959 and 1961 of the Mercury spacecraft in ballistic flight. Tested under various dynamic pressures were its escape system, stability, controls and recovery of the spacecraft and biological effects of acceleration and deceleration forces and weightlessness on rhesus monkeys.

In Little Joe, Mercury was boosted to about one fifth of orbital velocity. In contrast, on 9 September 1959, Big Joe using the Atlas launch vehicle that would launch Mercury, boosted an instrumented boilerplate model of Mercury to near orbital velocity to test entry into the atmosphere, the heat shield, flight characteristics and recovery. Results were so gratifying that no further tests were planned.

On 29 July 1960, MA-1 (Mercury Atlas, first), an unmanned suborbital flight with an Atlas booster rocket designed to check structural integrity under maximum heating conditions exploded after lift off. Investigations delayed Mercury for about six months. On 8 November 1960, Little Joe 5 failed when booster, tower, and capsule did not separate as planned. In the MR-1 test of 7 November 1960, the Redstone booster ignited and shut down. The escape tower then took off from the spacecraft. On 19 December 1960, NASA launched MR-1A. Both the Redstone and the Mercury spacecraft operated superbly. MR-2, with chimpanzee Ham as passenger, went higher and farther than planned on 31 January 1961, and both the spacecraft and its passenger were recovered in excellent condition. The Mercury-Atlas MA-2 unmanned flight test on Feb. 21, 1961, met every expectation. Little Joe 5A, 18 March 1961, did not go as planned but the Mercury-Redstone Booster-Development flight of 24 March 1961 indicated that all major booster problems were eliminated.

On 12 April 1961, the Soviet Union launched Vostok 1 with Major Yuri Gagarin aboard into earth orbit. After one orbit, he returned to earth. Cosmonaut Gagarin made history as the first human in space. On 25 April 1961, NASA launched the unmanned MA-3. The flight was aborted, but the escape system operated perfectly, enabling the spacecraft to be recovered. On 28 April 1961, Little Joe 5B demonstrated the ability of the Mercury escape system to function under the worst conditions a Mercury-Atlas launch would impose.

On 5 May 1961, Astronaut Shepard became America's first man in space. He rode his *Freedom 7* Mercury spacecraft on a 15-minute suborbital flight. From launch to landing, everything went as planned. America had its first man in space!

On 21 July 1961, Astronaut Grissom and his *Liberty Bell 7* completed the second Mercury-Redstone suborbital flight. After landing, the hatch cover on his spacecraft blew off, water poured in, and Grissom had to abandon ship. *Liberty Bell 7* was lost. On 13 September 1961, an unmanned Mercury was launched by Atlas into a one-orbit flight test. The test demonstrated that

***Below left:* A photo of Mercury capsule *Freedom 7* as it was being hoisted up the gantry to be mated to its Redstone booster for the historic Mercury Redstone-3 mission. *Below:* America's first man in space, Alan Shepard Jr. Astronaut Group 1—*at right,* clockwise from left rear: Alan Shepard Jr, Virgil Grissom, L Gordon Cooper, M Scott Carpenter, John Glenn Jr, Donald Slayton and Walter Schirra Jr. *Below right:* Mercury capsules in the assembly stage at McDonnell Douglas.**

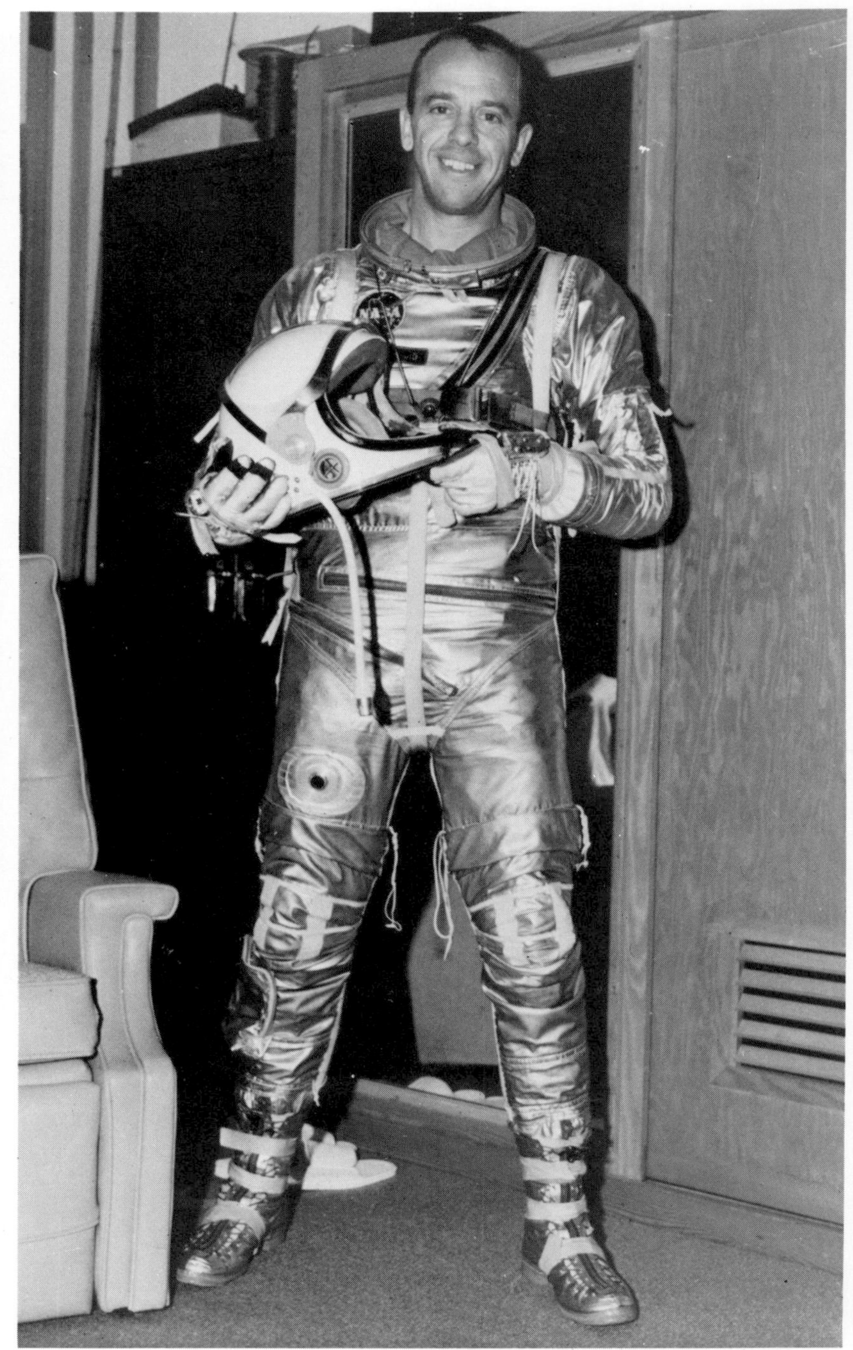

MAJOR ASSEMBLY
17
15A
20
12B
9A

a man could be safely orbited and returned to earth. On 29 November 1961, NASA launched Enos, a chimpanzee, into earth orbit.

Astronaut John Glenn was launched into earth orbit on 20 February 1962. He and his *Friendship 7* spacecraft gave Americans their first manned orbital flight, three orbits in nearly five hours. He took pictures and performed attitude-control and other chores during the flight. Glenn and *Friendship 7* achieved the primary objective of Mercury.

On 24 May 1962, Carpenter in the Mercury capsule named *Aurora 7* flew another three-orbit mission. His spacecraft was modified and the flight emphasized scientific rather than engineering experiments. He drifted for 77 minutes to conserve fuel and evaluate the effects of drifting on himself and the spacecraft. He overshot his landing point in the Pacific by 250 miles because of a yaw error but was picked up in about three hours. His craft was recovered later.

Launched 3 October 1962, in *Sigma 7*, Walter 'Wally' Schirra contributed to developing techniques applicable to extended space flight. In nearly every respect, from launch to recovery, his was a textbook mission. L Gordon 'Gordo' Cooper's *Faith 7* launched 15 May 1963, spent more than a day in space, the longest Mercury flight. He conducted earth photography and other scientific experiments. Cooper slept during rest periods, answering the question about whether sleep was possible during space flight.

John Glenn Jr became the first American to orbit the earth on 20 February 1962, during the Mercury Atlas-6 mission. Glenn is seen *at right* as he orbited the earth three times in his capsule, *Friendship 7*, which is seen atop its Atlas booster in the MA-6 takeoff photo at *far right*. Mercury Atlas-7 occurred on 24 May 1962. During MA-7, astronaut Scott Carpenter (seen *below*, peering into his *Aurora 7* capsule prior to climbing in) also completed three orbits of the earth.

President John F Kennedy, (Excerpts from his Address to the Congress 21 May 1961):

'Space is open to us now, and our eagerness to share its meaning is not governed by the efforts of others . . . we go into space because whatever mankind must undertake; free men must fully share.

'. . . I believe that this nation should commit itself to achieving the goal, before this decade is out, of landing a man on the moon and returning him safely to the earth. No single space project in this period will be more impressive to mankind or more important to the long-range exploration of space . . .

'In a very real sense it will not be one man going to the moon; if we make this judgement affirmatively, it will be an entire nation . . . if we are to go only half-way or reduce our sights in the face of difficulty, in my opinion it would be better to not go at all . . .

'. . . For while we cannot guarantee that one day we shall be first, we can guarantee that any failure to make this effort will surely make us last . . .'

The Gemini Program

Gemini markedly advanced the technological and experiential frontiers of manned space flight and vastly increased knowledge about space, earth and human biology. It demonstrated that pilots can maneuver a spacecraft; that properly equipped and clothed, they could live and work outside of their craft; that they could rendezvous and dock with another vehicle; that they could control their craft during descent from orbit; and that they could function effectively for flights up to two weeks in duration. It provided mastery of crucial technology and skills needed for Apollo, the moon-landing mission.

The photographs that the Gemini astronauts took of earth provided a wealth of information related to geography, resources and environment and demonstrated the value of a future earth survey system aided by satellites. Study of Gemini astronauts before, during and after flight and the medical electronic equipment developed for Gemini contributed to advances in instruments used for health care such as those that detect, measure and monitor life processes.

Gemini also contributed to astronomy and to studies of magnetic fields, radiation and micrometeoroids around earth.

Gemini was a three-part spacecraft — the reentry module where the crew lived and worked and the only part that returned to earth; the middle part was the adapter retrograde section that contained the retrorockets fired by the astronauts to leave orbit; and the third was the adapter equipment section, containing additional equipment and flaring out to mate with the Titan launch vehicle. The Gemini launch vehicle was a modified US Air Force Titan 2.

When the astronauts were ready for return to earth, they jettisoned the equipment section, fired the retrorockets of the retrograde section to slow down and descend from orbit, and then jettisoned the retrograded section.

Gemini 1, an unmanned test mission, lifted off on 8 April 1962 to determine performance and structural integrity of Gemini 2. After three orbits, all experiment objectives were achieved. No recovery was planned. The vehicle entered the atmosphere 12 April. Gemini 2 was also an unmanned suborbital test. Delayed by being struck by lightning, hurricane warnings and a misfire of its launch vehicle, it finally lifted off on 19 January 1965. It proved that the heat shield could protect Gemini under maximum heating during reentry and performed so well that it cleared the way for a manned launch.

Gemini 3, launched 23 March 1965, was named *Molly Brown* by its commander, Virgil 'Gus' Grissom. He and his pilot, John W Young, conducted the first manual maneuvers, changing not only the shape but also the plane of their orbit, a first for American spaceflight.

James A McDivitt and Edward H White II were the crew of Gemini 4 which circled earth from 3–7 June 1965. White accomplished the first American (EVA) extra-vehicular activity or spacewalk.

Gemini 5, launched 21 August 1965, circled the earth for eight days, a new world record. It demonstrated that the crew of L Gordon Cooper Jr and Charles Conrad Jr, and spacecraft were capable of prolonged functioning in space. No lasting harmful effects of apparent weightlessness were detected.

Gemini 7, launched 4 December 1965, confirmed in a 14-day flight that no lasting harm results from weightlessness. This broke the record of Gemini 5 for the longest manned flight. The crew was Frank Borman and James A Lovell Jr. Gemini 6, with Walter M Schirra Jr and Thomas P Stafford, was launched 15 December 1966, for its one-day flight. The launch was originally scheduled for 25 October, but was delayed because of problems with the launch vehicle. Gemini 6 successfully rendezvoused with Gemini 7, completing its primary mission — the world's first rendezvous of two vehicles in space — and conducted other experiments.

Gemini 8, with Neil A Armstrong and David R Scott at the controls, completed the world's first docking in space. However, after docked-vehicle maneuvers, the joined vehicles began an inexplicable spin. Armstrong undocked from the Agena target vehicle, but the Gemini spin continued. Finally, he fired the reentry rockets. This stopped the spin and cut the mission short. An emergency landing was made off Okinawa. The problem was an attitude-control thruster stuck in the firing position.

Gemini 9, on 3–6 June 1966, manned by Thomas P Stafford and Eugene A Cernan, performed a number of maneuvers with a target stage. It did not dock because the shroud covering the target's docking apparatus had failed to separate. Cernan made a two-hour spacewalk. Gemini 10, launched 18 July 1966, manned by John W Young and Michael Collins, docked with an Agena target vehicle and later rendezvoused with the Gemini 8 target vehicle that was still in orbit. It used the Agena's power to drive the joined vehicles to a higher altitude and to the Gemini 8 target vehicle. Gemini 10

Below: **Gemini 3's Virgil Grissom and John Young, with American artist Norman Rockwell and a NASA technician.** ***Above:*** **A Titan 2 booster hurtles Gemini 6 aloft.** ***Above right:*** **A Gemini 4 photo of Edward White II, as he assayed the first American spacewalk.** ***At right:*** **Earth's cloudtops as photographed by Gemini 12.**

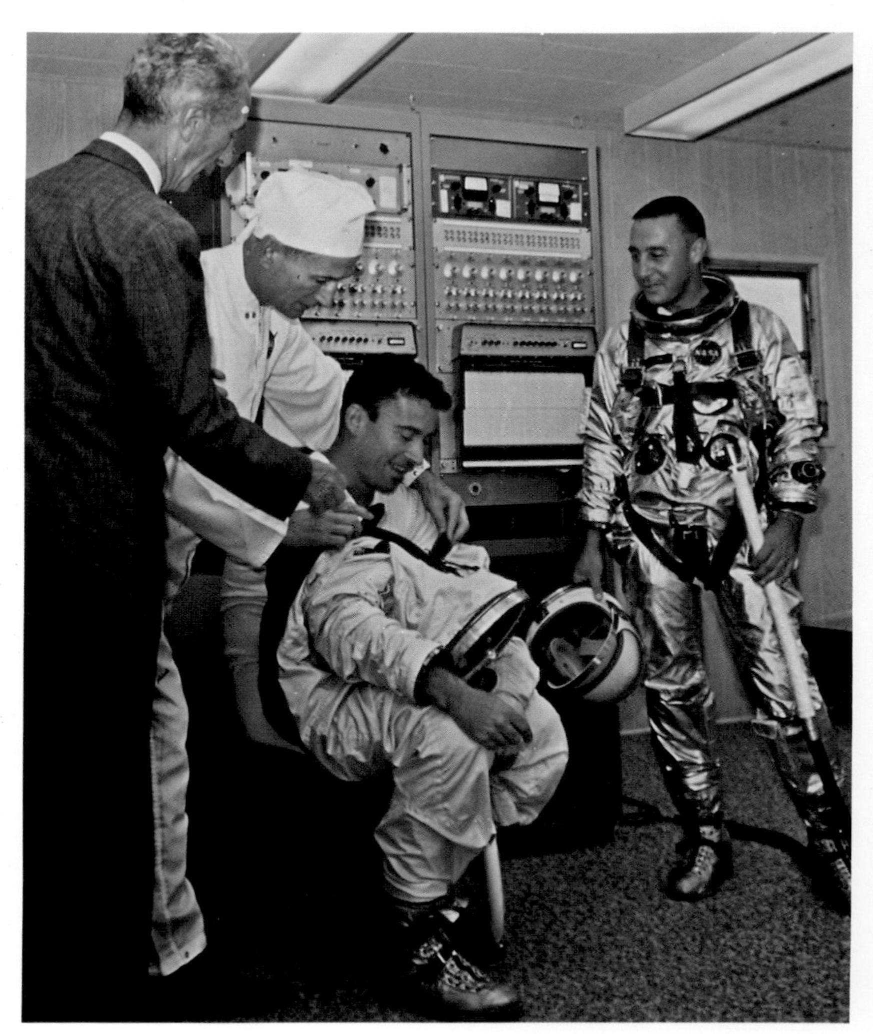

Gemini 7, with Frank Borman and James Lovell on board, was launched on 4 December 1965, and completed 206 orbits in 330.6 hours, breaking Gemini 5's record for the longest manned spaceflight. While in orbit, Gemini 7 served as a passive rendezvous target for Gemini 6 — launched on 15 December 1965, with its crew of Walter Schirra Jr and Thomas Stafford.

On these pages is a photograph taken by the Gemini 6 crew as their spacecraft closed to within six feet of Gemini 7, at which point the two craft maintained their positions for 5.5 hours.

flew in a formation with Agena while the two craft were linked by a tether. Collins did a stand-up EVA in the open hatch of Gemini and later used a handheld maneuvering unit to propel himself around. He landed on the Agena and retrieved a micrometeroroid experiment. Gemini 10 returned to earth on 21 July.

Gemini 11, on 12–15 September 1966, with Charles Conrad Jr and Richard F Gordon Jr, accomplished the world's first rendezvous and docking in the first orbit.

Using Agena rocket power, it soared to an apogee of 1372 kilometers, an altitude record for manned space fight. Gordon spent about two hours in a stand-up EVA and about a half hour floating and working outside of his craft.

Gemini 12, on 11–15 November 1966, was manned by James A Lovell Jr and Edwin E Aldrin Jr. Thoroughly trained in a simulated space environment on earth and helped by the latest advances in hand and footholds, Aldrin completed about 20 assigned EVA tasks in approximately two hours. His total EVA time including stand-up EVA and a walk in space was about five and one-half hours.

Closing in on the Moon

Even as Gemini progressed, unmanned Ranger, Surveyor and Lunar Orbiter spacecraft were scouting the moonscape for the Apollo program landings. They helped determine the bearing, strength and detailed topography of the moon and provided other data needed to plan moon landings.

Three successful Ranger flights—Ranger 7, launched 28 July 1964; Ranger 8, 17 February 1965; and Ranger 9, 21 March 1965—provided lunar photography that was as much as 2000 times as detailed as the best telescopic pictures. They showed that the seemingly smooth lunar plains were pockmarked with craters and that although the moon had no water, it was subject to erosion. Scientists attributed the latter to micrometeoroid bombardment. So impressed was the International Astronomical Union with Ranger 7 pictures that it renamed the dry Sea of Clouds area that Ranger photographed Mare Cognitum—Known Sea.

Lunar Orbiters helped identify suitable landing sites for Apollo and gave ground trackers experience which they would need for Apollo. They photographed nearly the entire lunar surface including the part that is always turned away from earth. They also furnished data about radiation and meteoroids in the moon's vicinity. They were deliberately crashed on the moon when they finished their mission so that they would not collide with Apollo or other future spacecraft.

Lunar Orbiter 1, launched 10 August 1966, was the first spacecraft to be placed in lunar orbit. It was also the first to photograph the earth from the moon. Lunar Orbiter 2 was launched 6 November 1966; Lunar Orbiter 3, 5 February 1967; Lunar Orbiter 4, 4 May 1967; and Lunar Orbiter 5, 1 August 1967. All were successful.

Surveyor 1, launched 30 May 1966, soft landed in the dry Ocean of Storms. It found the bearing strength of the lunar surface was more than adequate to support the Apollo landing craft. This contradicted a theory that the lunar landing craft might sink into an ocean of dust. Surveyor also telecast many pictures to earth. Surveyor 3, launched 17 April 1967, landed

At right: **One of the Lockheed Agena target vehicles used in NASA's early space docking experiments.** ***Below:*** **Gemini 10 astronauts (left and right, respectively) John Young and Michael Collins return from their July 1966 docking and EVA mission (see text, pages 118–this page).** ***Below right:*** **A Ranger spacecraft mockup.**

in another area of the Ocean of Storms. Armed with a shovel, it dug a trench and found that bearing strength increased with depth. It also sent pictures. Surveyor 5, launched 8 September 1967, landed in the dry Sea of Tranquility. Its alpha scattering device, probing the chemical composition of the surface, revealed a similarity to basalt on earth.

Surveyor 6, launched 7 November 1967, landed in the moon's Central Bay. In an extremely significant experiment, NASA engineers remotely fired Surveyor's vernier rockets to launch it briefly above the surface. Surveyor's launch raised no dust cloud and resulted in only shallow cratering. Surveyor 6 also sent thousands of pictures and many hours of alpha scattering data. Surveyor 7, 7 January 1968, landed in a highland area of the moon, near the Crater Tycho. Its alpha scattering device showed that the highlands soil contained less iron than plains soil. It photographed laser beams from earth, a significant communications test.

The builders of the Apollo spacecraft needed information about micrometeoroids for design purposes. The first satellite intended specifically for meteoroid studies was Explorer 16, launched 16 December 1962. Its instrumentation was built around the spent upper stage of the Scout launch vehicle. Similar meteroid Explorers were Explorer 23, launched 6 November 1964, and Explorer 46, launched 13 August 1972. They tested abilities of various kinds of structures to resist micrometeoroid punctures and reported on penetration frequencies.

Three Pegasus satellites, which were so huge that they could be seen in the night sky by the unaided eye on earth, were launched on 16 February, 25 May and 30 July 1965. At launch, Pegasus consisted of the Saturn IB launch vehicle, predecessor of the uprated Saturn IB used in Apollo manned earth orbital flights; boilerplate Apollo Command and Service Modules; and the Apollo launch escape system. The launch escape system was jettisoned during launch as part of an Apollo test. Data on Saturn 1B performance during launch contributed to development of Saturn 1B and to Saturn 5, which would launch Apollo to the moon.

After separating in orbit from the boilerplate Command and Service Module, Pegasus unfolded 96-foot long wings designed to expose a broad range of thicknesses to micrometeoroids. One result of Pegasus data was discovery of lower meteoroid density than expected, enabling planners to reduce Apollo weight by about 1000 pounds.

Below: A Lunar Orbiter model. _At right:_ An Atlas Centaur booster launches Surveyor 1 toward the moon. _Below right:_ A Surveyor spacecraft model. Surveyor 3 touched down on the moon in April 1967, and manned lunar mission Apollo 12 followed suit in November 1969 (see page 132). _At far right:_ An Apollo 12 mission photo with Surveyor 3 in the foreground, and the Apollo 12 Lunar Module in the background.

The Apollo Program

For centuries, humanity had fantasized about travelling to the moon. NASA's Apollo program turned fantasy into reality. Americans walked on the moon's ancient untrodden surface. They unveiled secrets that had eluded humanity since it first began to think about the moon.

The first Apollo expedition landed on the moon a little more than eight years after the national commitment was made. In 1961, when President Kennedy had vowed to have Americans on the moon by the end of the decade, American astronauts were still flying the one-man Mercury spacecraft. Indeed, only one American had been in space when Kennedy articulated the goal. The lunar landing reflected giant steps forward.

Within a year after a lunar landing commitment had been made, development of the powerful boosters — Saturn 1B, needed for earth orbital tests, and Saturn 5, needed for the lunar launch was in progress. The lunar orbit rendezvous procedure called for Apollo to be launched from earth into lunar orbit and a landing craft to be detached to carry two humans to the lunar surface. Later, the pair would rocket their craft from the moon and rendezvous and dock with the orbiting parent craft. The landing craft was called the Lunar Module. The craft in which the crew would ride to the moon and back was the Command Module. A third section was the Service Module containing the main propulsion system and supplies of water and air. It would remain attached to the Command Module. Only the Command Module returned to earth. The Service Module would be jettisoned just before reentry.

Apollo progressed rapidly. By 1965, Saturn 1B had successfully completed its six-launch program in which it not only demonstrated the practicability of the clustered-engine concept and a liquid-hydrogen second stage, but also tested a boilerplate Apollo spacecraft and launched three Pegasus satellites. Tests with the uprated Saturn IB and Apollo Command and Service Modules were completed in 1966. Then Apollo Command and Service Modules were considered ready for manned earth orbital flight tests.

On 27 January 1967, as astronauts Gus Grissom, Edward H White II and Roger B Chaffee were conducting a countdown toward a simulated launch in the Apollo 1 Command Module on the launch pad at Cape Kennedy, Florida, when tragedy struck. Fire broke out in the spacecraft. In the approximately five minutes it took for rescuers to open the hatch from the outside, the three astronauts had died from asphyxiation. It was the first fatal accident of the space program. A stunned nation mourned.

The Apollo Command Module was redesigned and rebuilt. Among the changes were a different hatch, reworked wiring and noncombustible materials. Finally, on 9 November 1967, NASA launched the unmanned Saturn 5/Apollo 4 into orbit. This test of the launch vehicle and Apollo's ability to reenter the atmosphere at lunar-return speed (about 25,000 mph) provided provided satisfactory results. On 22 January 1968, in Apollo 5, a Saturn IB launched the unmanned Lunar Module into earth orbit. Lunar Module ascent and descent engines were fired twice in orbit.

Saturn 5/Apollo 6 was launched in another unmanned test on 4 April 1968. Saturn 5 exhibited some unwanted characteristics such as oscillations called POGO but was considered safe for manned space flight. The wisdom of this decision was confirmed with the earth orbital mission of Walter M Schirra, Donn Eisele and Walter Cunningham on 11-22 October 1968. The three-man Apollo spacecraft completed a successful earth-orbital mission of 11 days.

On 21 December 1968, Frank Borman, James A Lovell Jr and William Anders flew Apollo 8 to the moon, 10 times around the moon and returned

At right: **An original 'block one' Apollo capsule, as delivered in 1966. Design changes were to result from the Apollo 1 fire.** ***Below:*** **The charred capsule in which the space program's first human fatalities occurred.** ***Below left:*** **The Apollo 1 crew, from the top: Virgil Grissom, Edward White II and Roger Chaffee.**

UNITED
STATES

safely to earth. Never before had people traveled so far, so fast or looked so closely at another celestial body. They were the first to view the backside of the moon which cannot be seen from earth. Arriving in lunar orbit on Christmas Eve, they returned a wealth of lunar pictures and verified the suitability of the lunar landing sites. Apollo 8 demonstrated that the Apollo Command and Service Modules operated as they were supposed to. The crew and their Apollo Command Module returned safely to earth 27 December.

Apollo 9, 3-13 March 1969, provided the first manned orbital flight test of the Lunar Module. This was the first manned flight of the complete Apollo spacecraft. Astronauts James A McDivitt, David R Scott and Russell L Schweickart, after separating their Command/Service Module from the Saturn 5 third stage, turned around and docked with the Lunar Module. They pulled it clear of the third stage. Later, McDivitt and Schweikart entered the Lunar Module, separated it from the Command Module, conducted a variety of maneuvers near and far from the Command Module, including simulated lunar descent and ascent, and finally rejoined the Command Module. Schweikart also conducted the first Apollo space walk. Call signs used for communication while the Command/Service Modules (CSM) and Lunar Module (LM) were separated were *Gumdrop* and *Spider*, respectively.

Apollo 10, 18–26 May 1969, conducted America's second manned circumlunar flight. Thomas P Stafford and Eugene A Cernan separated the Lunar Module, *Snoopy*, from the CSM *Charlie Brown*. John W Young continued to pilot *Charlie Brown* around the moon. Stafford and Cernan conducted what has been called a dress rehearsal for the lunar landing. They maneuvered Snoopy to practice rendezvous and docking, descent to as low as nine miles above the moon, and ascent.

Below: **Apollo 9's Lunar Landing Test Module *Spider* in earth orbit. Note the surface probes extending from *Spider's* legs. This earth orbital test helped to prepare for the Apollo 11 manned lunar landing mission. *At right:* The Apollo 11 Command Service Module is here being mated to its Saturn 5 launch vehicle. *At far right:* Apollo 11 blasts off at Kennedy Space Center's Pad 39A on 16 July 1969.**

On 20 July 1969, the Apollo Lunar Module *Eagle* landed in the moon's Sea of Tranquility, 102.8 hours after leaving the earth. Mission commander Neil Armstrong opened the Lunar Module's hatch 6.4 hours later, and made his way down the *Eagle's* ladder to the surface of the moon. He kept up a running commentary as he made his historic descent: 'I'm at the foot of the ladder . . . the surface appears to be very, very fine grained . . . I'm going to step off the LM now. That's one small step for man . . . one giant leap for mankind.' *Above:* The first human foot and footprint on the moon. *At left:* Twenty minutes after Neil Armstrong, Edwin Aldrin descends the Lunar Module's ladder to become the second man on the moon. Not one to be outdone in the realm of poetics, his quote was 'Magnificent desolation.' Armstrong and Aldrin gathered 44 pounds of rock samples and conducted various other scientific investigations. *At right,* Edwin Aldrin sets up a solar wind experiment near the LM. *Below right:* Armstrong and Aldrin set up the United States flag. *Below left:* Neil Armstrong, in the Lunar Module at Tranquility Base after his EVA. *Below:* Edwin Aldrin, in the LM after EVA. Michael Collins piloted, and remained in lunar orbit with, the Command/Service Module *Columbia.*

Apollo 11 (16-24 July 1969) **Astronauts Neil A Armstrong, Buzz Aldrin, Michael Collins.**
(First words from Tranquility Base, first manned lunar landing, 5:18 pm EDT, 20 July 1969):
Capcom: (Charles Duke) 'We copy you down Eagle.'
Eagle: 'Tranquility base here, the Eagle has landed.'
Capcom: 'Tranquility, we copy you on the ground. You've got a bunch of guys about to turn blue. We're breathing again, thanks a lot.'

'That's one small step for a man; one giant leap for mankind.' These were the words of Neil A Armstrong, commander of Apollo 11, as he set foot on the moon at 10:56 pm EDT, 20 July 1969. He and his Lunar Module pilot, Edwin E 'Buzz' Aldrin Jr had landed the LM *Eagle* on the moon at 4:18 p.m. EDT, 20 July, while Michael Collins orbited overhead in the CSM *Columbia*. Their flight began on July 16 and they returned home to a cheering world on July 24. History had been made. Armstrong and Aldrin had landed, worked on and explored a part of the moon's dry Sea of Tranquility and returned safely to earth.

Apollo 12, 14–24 November 1969, was devoted to extensive exploration of another part of the moon – the dry Ocean of Storms. The surface and subsurface material that Charles Conrad Jr and Alan L Bean took home from the moon was different from that brought back in Apollo 11. Among the items brought back were parts of the Surveyor 3 unmanned lander. In addition, Conrad and Bean set up a sophisticated set of experiments called the Apollo Lunar Surface Experiments Package. Among its instruments, which would keep sending information to earth long after the Apollo expedition left, were a seismometer, magnetometer and other geophysical equipment. It was nuclear-electric powered. The Lunar Module's code name was *Intrepid*. Richard F Gordon Jr orbited overhead and conducted experiments in the Command Module *Yankee Clipper*.

Apollo 13, 11–17 April 1970, was a sobering reminder that the problems and dangers of space exploration are frequently beyond anticipation. Although carefully checked beforehand, the oxygen tank of the Apollo 13 Service Module ruptured while Apollo was enroute to the moon. Without the Service Module, the Command Module did not have the water, oxygen and propulsive power needed for a safe return.

Both the crew — James A Lovell Jr, Fred W Haise Jr, and John L 'Rusty' Swigert Jr — and ground controllers recognized that the unused Lunar Module could provide what they needed to get home. The crew lived in the module as the combined craft continued on its trajectory around the moon. They used the Lunar Module rockets to correct their course and align their craft properly for safe entry into earth's atmosphere, and abandoned the Lunar Module only when the Command Module was on course for a proper atmosphere entry. The crew and craft returned safely.

Apollo 14, in service from 31 January to 9 February 1971, was modified to prevent a recurrence of the Apollo 13 problem. Alan B Shepard and Edgar D Mitchell landed their Lunar Module *Antares* at Fra Mauro upland or foothill region different from the lunar plain which Apollo 11 and 12 had explored. Stuart A Roosa conducted experiments from the Command/Service Module *Kitty Hawk*. The work of Mitchell and Shepard was facilitated by use of a Modularized Equipment Transporter, a cart-like device. They set up another Apollo Lunar Surface Experiments Package, collected rock and dust samples from various places as they surveyed the area around Fra Mauro and conducted other scientific tasks.

The first motor trip on the moon was during the Apollo 15 mission. On 31 July 1971, David R Scott and James B Irwin removed their folded Lunar Roving Vehicle from the Lunar Module *Falcon* and unfolded and set it up for business. In three surface EVAs, they drove far and wide around the Hadley-Apennine region of the moon. They took photographs and rock samples, drove a core tube into the soil, set up the Apollo Lunar Surface Experiment Package and a laser reflector, and did other tasks. The Lunar Roving Vehicle camera telecast *Falcon*'s blast-off from the moon, which in

Below: **Apollo 12 Lunar Module *Intrepid* descends toward the moon's Ocean of Storms. *At far right:* Visible damage, caused by an exploded oxygen canister, on Apollo 13's Service Module. *At right:* Apollo 13 CM pilot John Swigert Jr holds an air scrubbing device that aided the Apollo 13 crew's survival. *Below right:* Apollo 14 astronaut Alan Shepard shades his eyes against the sunlight on the moon.**

These pages: Apollo 16 Lunar mission commander John Young leaps from the lunar surface as he salutes the United States flag near the Apollo 16 landing site at the crater Descartes. Lunar Module *Orion* is on the left here, and the Lunar Roving Vehicle is parked beside the LM. Lunar Module pilot Charles Duke took this photo. Thomas Mattingly piloted the Apollo 16 Command and Service Module *Casper*. This was the second Apollo mission to carry the Lunar Rover. Altogether, 20.2 hours of lunar EVA was logged by the LM crew. An idea of how cramped the Apollo astronauts were during their lunar odysseys, the Apollo Command Module, in which three astronauts per mission had to ride to and from the moon, offered 73 cubic feet of space per man, as compared with the 68 cubic feet per person allowed by the average compact car.

the space vacuum was evidenced as a shower of colorful sparks. After their return to the orbiting CSM *Endeavour*, piloted by Alfred M Worden Jr, a small subsatellite instrumented to study the moon for a year was launched. Enroute to earth, Worden conducted the first EVA in deep space as he retrieved film from a camera in the Service Module.

The Descartes area of the moon was explored by John W Young and Charles M Duke Jr as part of the Apollo 16 mission, 16–27 April 1972. The biggest surprise of the landing mission was that the Cayley Plain which from orbit looked smooth to undulating, suggesting past volcanic flow and igneous rock, was made up of breccias which are composed of rock fragments welded together by intense heat. Young and Duke, who landed in *Orion* used a Lunar Roving Vehicle and set up an Apollo Lunar Surface Experiment Package as did the Apollo 15 expedition. Thomas K Mattingly II operated a battery of cameras and other experiments as he orbited overhead in Casper. Enroute to earth, Mattingly went outside the spacecraft to retrieve film from the Service Module.

Apollo 17, on 7–19 December 1972, was the last and most productive of the lunar missions. Eugene A Cernan and Harrison H Schmitt Jr landed on the Taurus-Littrow area of the moon in *Challenger* while Ronald E Evans orbited in the CSM *America*. Using the Lunar Roving Vehicle, they traveled farther, spent more time and collected a heavier total of rock samples than in any previous mission. Their total mission and lunar orbit times were the longest of all. Because this was the last lunar expedition, man's last close view of the moon for years to come, the crew sought to see and collect all the data, material and photographs that they could.

The landing of Apollo 17 marked the end of an era. Knowledge about the moon and with it, earth, was expanded vastly by Apollo. Man's ability to perform useful tasks in an alien environment was demonstrated.

Hundreds of scientists in the United States and abroad participated in studying the lunar samples, photographs and other data brought back by Apollo expeditions. The interest in Apollo and in the Apollo astronauts transcended world boundaries. The United States, for its part, took the position that Apollo was a triumph of humanity even though the deeds were performed by Americans. The technology of Apollo has contributed to many fields such as development of instruments to detect, measure or monitor life processes or to improve production or promote safety in industry.

The greatest achievement of Apollo may be the relatively short time in which skills were established, machines and facilities built, and people and equipment organized on a vast scale to extend the limits of human activities beyond this planet to the moon.

The end of the Apollo era coincided with new views about how to achieve excellence in American life, a change in public priorities. Americans became more concerned with programs that would result in immediate tangible benefits.

Skylab, America's First Space Station

Having achieved an incredible series of six lunar landings, NASA's next manned spaceflight project was Skylab, America's first experimental manned space station. The goals of skylab were to:

- Evaluate systems and techniques for acquiring information about earth's resources and environment.
- Increase knowledge of the sun and solar-terrestrial relationships.
- Increase knowledge about effects of prolonged space flight and with it biomedical knowledge.
- Test industrial processes to which space vacuum and so-called space weightlessness can contribute.

Skylab used existing technology and hardware from the Apollo program. The Skylab workshop where the astronauts lived and worked was an empty third stage of Saturn 5 that launched Apollo to the moon. It was modified as a two-story building, one with a laboratory and the other with living quarters. Attached to the workshop were an airlock module, a multiple docking adapter and an Apollo Telescope Mount. It was about the size of an average three-bedroom house. It was the largest object America had placed in orbit.

A Saturn 5, such as was used to launch the lunar missions, was designated to place the huge 84-ton Skylab module in orbit. Then, a Saturn 1B, like those used in Apollo earth-orbital tests, orbited an Apollo Command/Service Module with a crew of three. The astronauts docked with the workshop, entered it and set up housekeeping and went to work. When

Lunar Roving Vehicles were deployed only on the last three Apollo missions, and traveled distances of 17.5 miles (Apollo 15), 16.6 miles (Apollo 16) and 22 miles (Apollo 17). *At right:* An Apollo astronaut and an LRV on the surface of the moon.

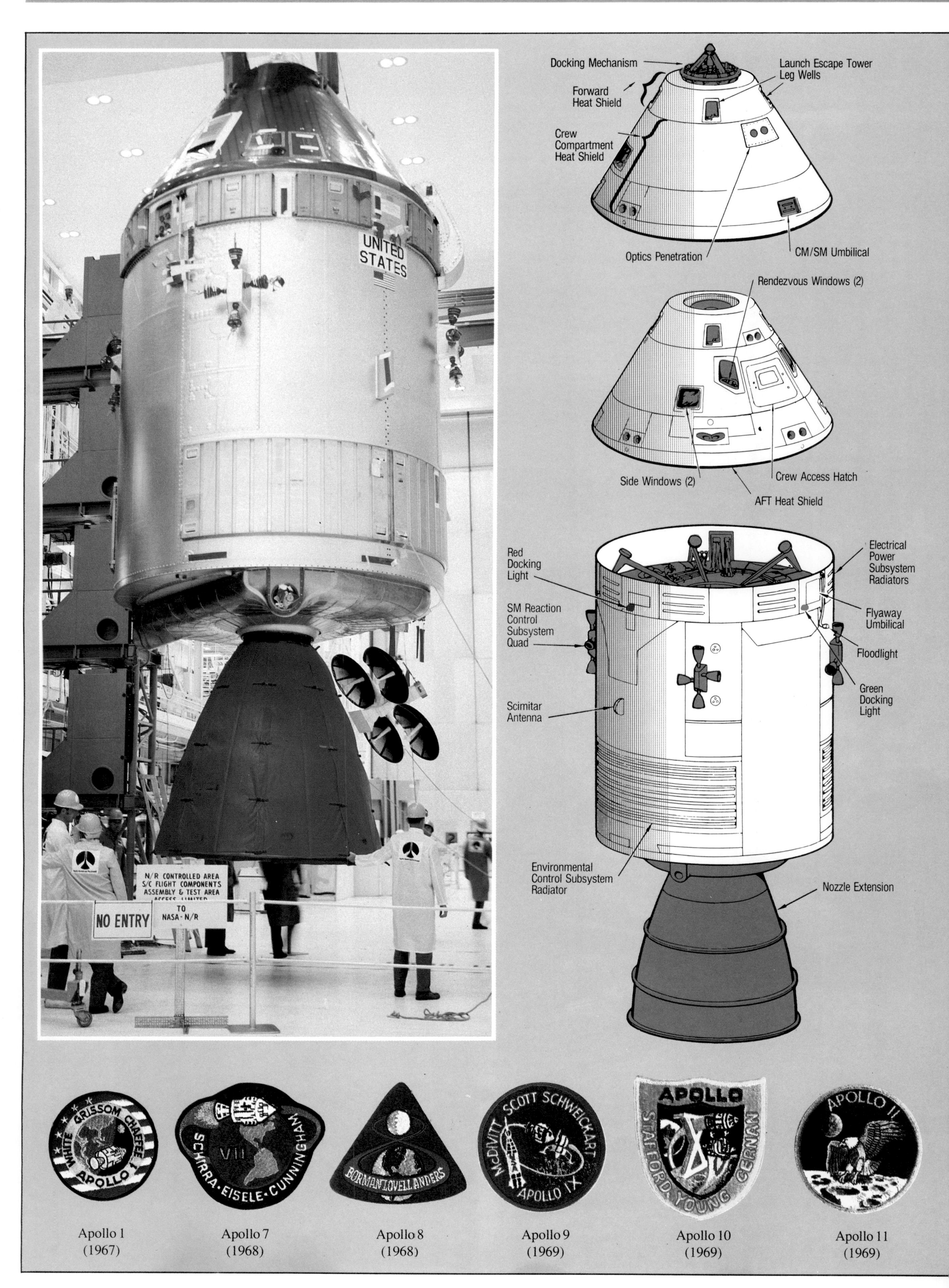

Apollo 1 (1967)

Apollo 7 (1968)

Apollo 8 (1968)

Apollo 9 (1969)

Apollo 10 (1969)

Apollo 11 (1969)

The Apollo Spacecraft

The hardware that supported the Apollo lunar missions between 1968 and 1972 was a system of some of the most sophisticated spacecraft ever designed. The centerpiece of the system was the Command/Service Module or CSM *(facing page)*. The photo shows the Apollo 13 CSM (CSM-109) freshly delivered to Cape Kennedy from Rockwell International, with blue protective coating on the Command Module and special red padding on the Service Module's rocket engines. The diagram *on the facing page* shows (from top) the Command Module's outer shield, the Command Module (CM) itself and the Service Module (SM). The overall diameter of the CSM was 12 feet, 10 inches. It stood four stories high and weighed six tons.

The top of the Lunar Module (LM), seen *below*, was a hatch which connected to the tip of the CM during the flight to the Moon. Having arrived in lunar orbit, two of the three astronauts aboard the CSM would crawl through the hatch, detach the LM and land on the Moon as shown in the photo *above*. Having completed their lunar excursion, the two astronauts would lift off from the Moon's surface in the LM's Ascent Stage only. The Ascent Stage held crew quarters and all the LM's control and communications systems as well as the ascent rocket engine. The Descent Module (DM), with foot pads, contained the motor used in the descent to the Moon's surface and was always abandoned there by the astronauts. The Ascent Module was, in turn, abandoned in lunar orbit after returning the two astronauts to the CSM. The entire crew would return to Earth orbit in the CSM, abandon the SM in orbit, and would return to Earth in the CM only.

Lunar Rovers *(in the photo above)* were carried aboard the LMs of Apollo 15, 16 and 17 in 1971 and 1972. Rated at a maximum speed of 8.7 mph, one was actually driven (downhill) at a breathtaking 10.6 mph. They weighed 462 pounds and measured 72 x 137 inches. The three Rovers that were driven on the lunar surface logged a total of 56.1 miles.

Each Apollo spacecraft was launched into space by one of two launch vehicles. The Saturn 1B *(right)* was used for missions in Earth orbit, while the huge Saturn 5 *(far right)* was used for lunar missions. The Saturn 1B, whose rocket engines delivered two million pounds of thrust, stood 224 feet high. The Saturn 5, whose engines delivered *nine million* pounds of thrust, stood 363 feet high. The Saturn 5 still stands as the most powerful rocket ever used for space flight.

The badges shown *across the bottom of this spread* represent all Apollo lunar landing program missions with crews assigned. All went into space except Apollo 1, which was destroyed by fire. All others went to the Moon except Apollo 7 and 9, which were designated for Earth-orbit missions only. All missions from Apollo 11 forward landed on the Moon except Apollo 13, which aborted its mission en route because of an explosion in the SM. Apollo 17 was followed by four Apollo missions unrelated to lunar landing missions.

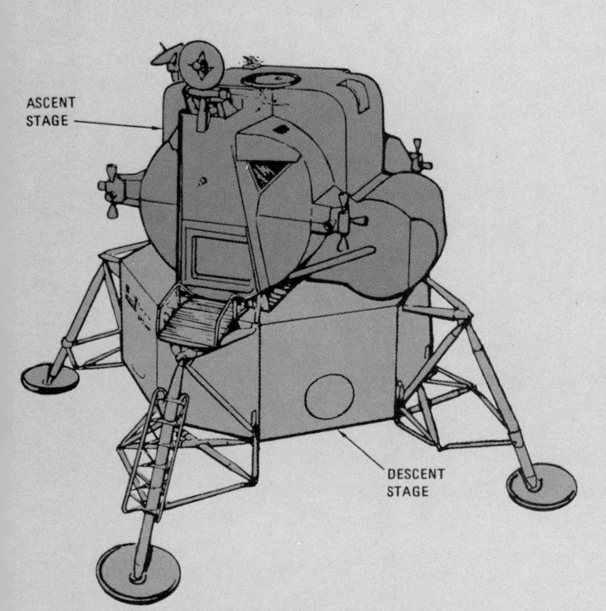

Apollo 12 (1969) · Apollo 13 (1970) · Apollo 14 (1971) · Apollo 15 (1971) · Apollo 16 (1972) · Apollo 17 (1972)

the mission was completed, the astronauts returned to their Apollo Command/Service Module, undocked and returned to earth.

Skylab was launched on 14 May 1973. Soon after, however, telemetry told ground controllers that the meteroid shield, needed for protection against both meteroids and solar heat, had been ripped off during the launch. In flying off, the shield tore off one solar panel and jammed the other, leaving the workshop without most of its electrical power and with the prospect of baking in the sun.

Engineers worked to devise a makeshift deployable shield and special tools for freeing the solar panel and other vital tasks. The first Skylab crew—Charles Conrad Jr, Joseph P Kerwin and Paul Weitz, carried these and other supplies into space when they were launched on 25 May 1973. After docking, the crew entered the workshop and deployed the shield. The sizzling temperature inside began to fall and in about 11 days had reached a comfortable 75 degrees Fahrenheit.

On 7 June, Conrad and Kerwin in a four-hour operation succeeded in freeing the solar panel. The surge of power began to charge eight batteries and provide about 3000 watts of desperately needed electricity.

In addition, the crew fixed a number of other malfunctioning instruments. Among them were a jammed gear mechanism for driving an ultraviolet telescope, a balky tape recorder, a battery not producing power because of a stuck contact in its regulator (Conrad hit it with a hammer, a classic fixit technique) and non-operating valves in the cooling system. Thus human intelligence and ability to meet unanticipated situations enabled Skylab to fulfill its mission.

Conrad, Kerwin and Weitz worked for 28 days on this space mission. They obtained thousands of earth survey and solar pictures, exceeding the wildest expectations of solar physicists, representatives of industry, agriculture, and weather services and city planners, ecologists, fisherman, prospectors and others concerned with earth's resources and environment. The crew's ability to work in space was much greater than anticipated.

The second Skylab crew—Alan L Bean, Owen K Garriott and Jack Lousma—more than doubled the time in space of the first crew. During a 59-day mission beginning 25 May 1973, they obtained 77,600 telescope images of the sun's corona the study of which vastly increased knowledge about solar processes. They also obtained 14,400 earth-survey pictures which were subsequently used in 116 investigations, both in the United States and abroad, in nearly every field of earth sciences. They formed crystals and metallic spheres more perfect and alloys stronger than those made on earth. They conducted experiments with spiders (Arabella and Anita) and found that, after a period of adaptation, spider webs spun in space resembled those on earth.

They also observed a mystifying development with Mummichug minnows. Those brought aboard swam at first in a disoriented manner. Those hatched from eggs brought aboard swam normally.

Blood and other samples the crew brought back with them added to knowledge about biochemical changes in space and contributed to medical

At right: **The Skylab launch from Pad 39A on 14 May 1973, using a modified Saturn 5 booster.** ***Below:*** **The Skylab space station, as seen from the airlock/docking end. At upper right is the Apollo Telescope Mount with its four solar panels.**

USA

knowledge of life processes on earth. Doctors discovered that exercise could stabilize or slow down deleterious effects of weightlessness.

The astronauts also performed engineering work aboard their craft such as installing a new parasol over the first one and plugging in a new assembly of gyroscopes to keep Skylab steady.

In the final Skylab mission, Gerald P Carr, Edward G Gibson and William R Pogue spent 84 days between launch on 16 November 1973, and landing 8 February 1974. They spent more time on major medical experiments than either of the two other crews, contributing to knowledge of the human body as well as to planning for future long duration missions. The record of Skylab indicated that with proper exercise and diet, there may be no limit to how long humans can make their home in space. With proper hand and footholds and tools, there is also no apparent limit on the work that one can do in space.

In addition to the longest flight time, the last group of Skylab astronauts spent more than 22 hours in EVA, another new record.

Skylab clearly demonstrated that man can function effectively without harmful after effects for prolonged periods in space. It produced an unprecedented wealth of data in such diverse fields as study of the sun, medicine, industrial processes in space and earth surveys.

It was originally hoped that another group of astronauts using the Space Shuttle would visit Skylab in the future, retrieve a time capsule bag to learn about reaction of articles to long term space exposure, and perhaps drive Skylab to a higher and longer term orbit. However, solar activity increased atmospheric density at Skylab's altitude, causing drag on the craft and decay of its orbit. On 11 July 1979, Skylab was destroyed by re-entry into the atmosphere.

At right: During EVA on the second Skylab manned mission (Skylab 3), astronaut Jack Lousma's helmet visor reflects Skylab and planet earth. _Below:_ Skylab in orbit, during Skylab 4. Note the missing left side solar panel, torn off at launch. _At far right:_ McDonnell Douglas engineers conduct usage tests in Skylab.

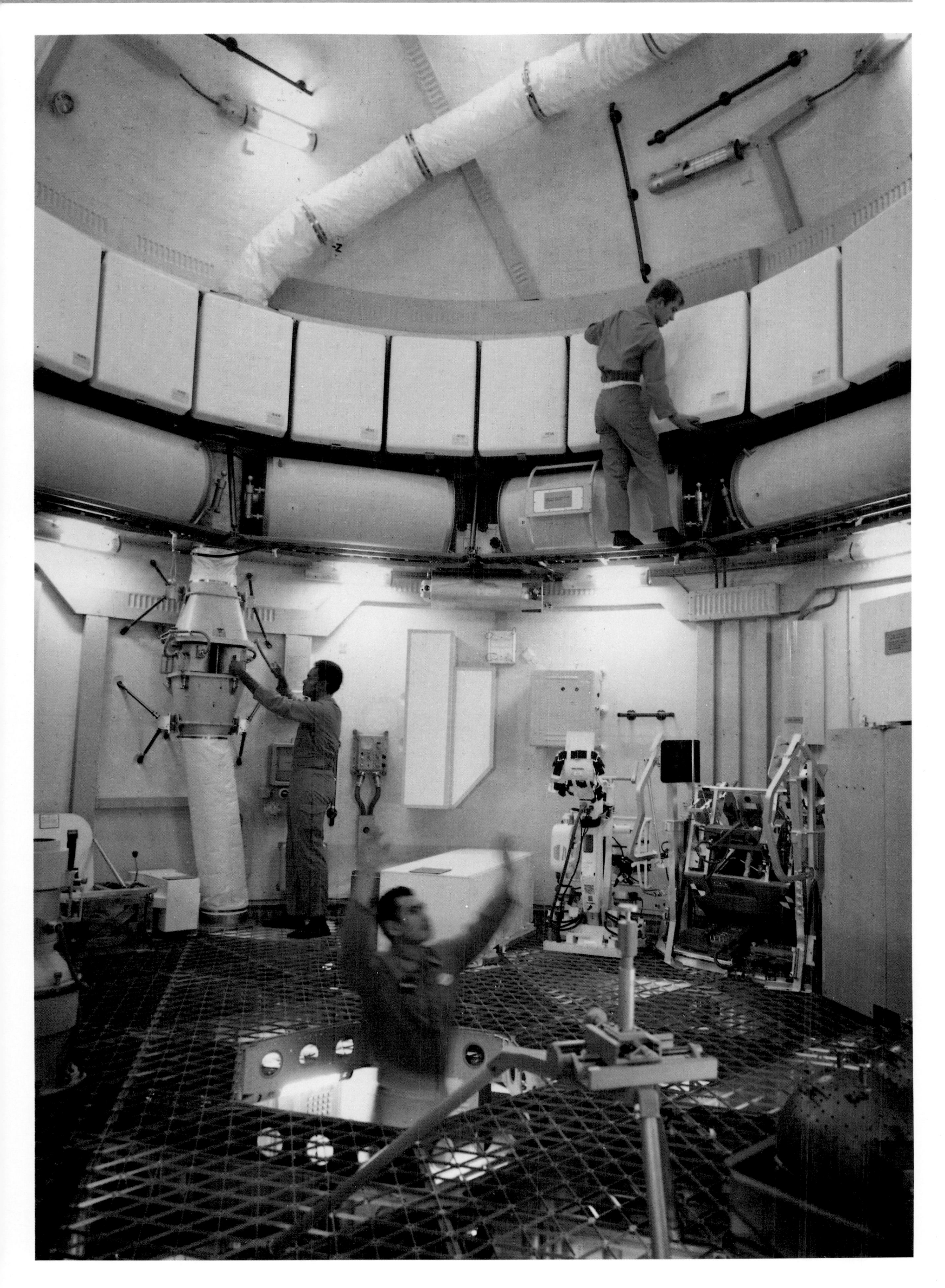

The Apollo Soyuz Test Project

In July 1975, three American astronauts and two Russian cosmonauts docked their two spacecraft in earth orbit, exchanged visits, and conducted joint and independent scientific and technical experiments. The principal goal of this mission, termed the Apollo Soyuz Test Project (ASTP), was to test compatible rendezvous and docking systems for manned spacecraft. The systems worked, opening the way for international space rescue, if necessary, and for future international manned space missions.

Both Apollo and Soyuz were launched on 15 July 1975. Soyuz, with cosmonauts Aleksey A Leonov and Valeriy N Kubasov aboard, was launched from the Baikonur Cosmodrome near Tyuratam in the Kazakh Soviet Socialist Republic at 8:20 am EDT. Apollo, with astronauts Thomas P Stafford, Vance D Brand and Donald K Slayton aboard, was launched by a Saturn 1B from Kennedy Space Center, Florida, at 3:50 pm EDT. The world's first international manned space flight rendezvous and docking were completed at 12:12 p.m. EDT, 17 July. The world's first international handshake in space, between Stafford and Leonov, took place at 3:19 pm EDT.

Apollo and Soyuz docked twice during the mission. Final undocking was at 11:26 am EDT, on 19 July. The Soyuz Descent Vehicle landed in Kazakhstan at 6:51 am EDT, on 21 July. The Apollo Command Module landed at 5:18 pm EDT, on 24 July, in the Pacific Ocean, as planned, where the craft and crew were picked up by a waiting recovery force.

Apollo-Soyuz involved 28 separate scientific and technical experiments. Areas covered were astronomy, earth observations, life sciences, and a variety of applications including such tests as crystal growth and electrophoresis in space.

At right: **The Apollo Soyuz Test Project (ASTP) launch on 15 July 1975.** ***Below:*** **The Soyuz spacecraft, as seen from Apollo during the ASTP.** ***At far right:*** **Apollo Docking Module (DM) pilot Donald Slayton and Soyuz commander Alexei Leonov in ASTP onboard activity.** ***Below far right:*** **Apollo, with ASTP DM, photographed by Soyuz.**

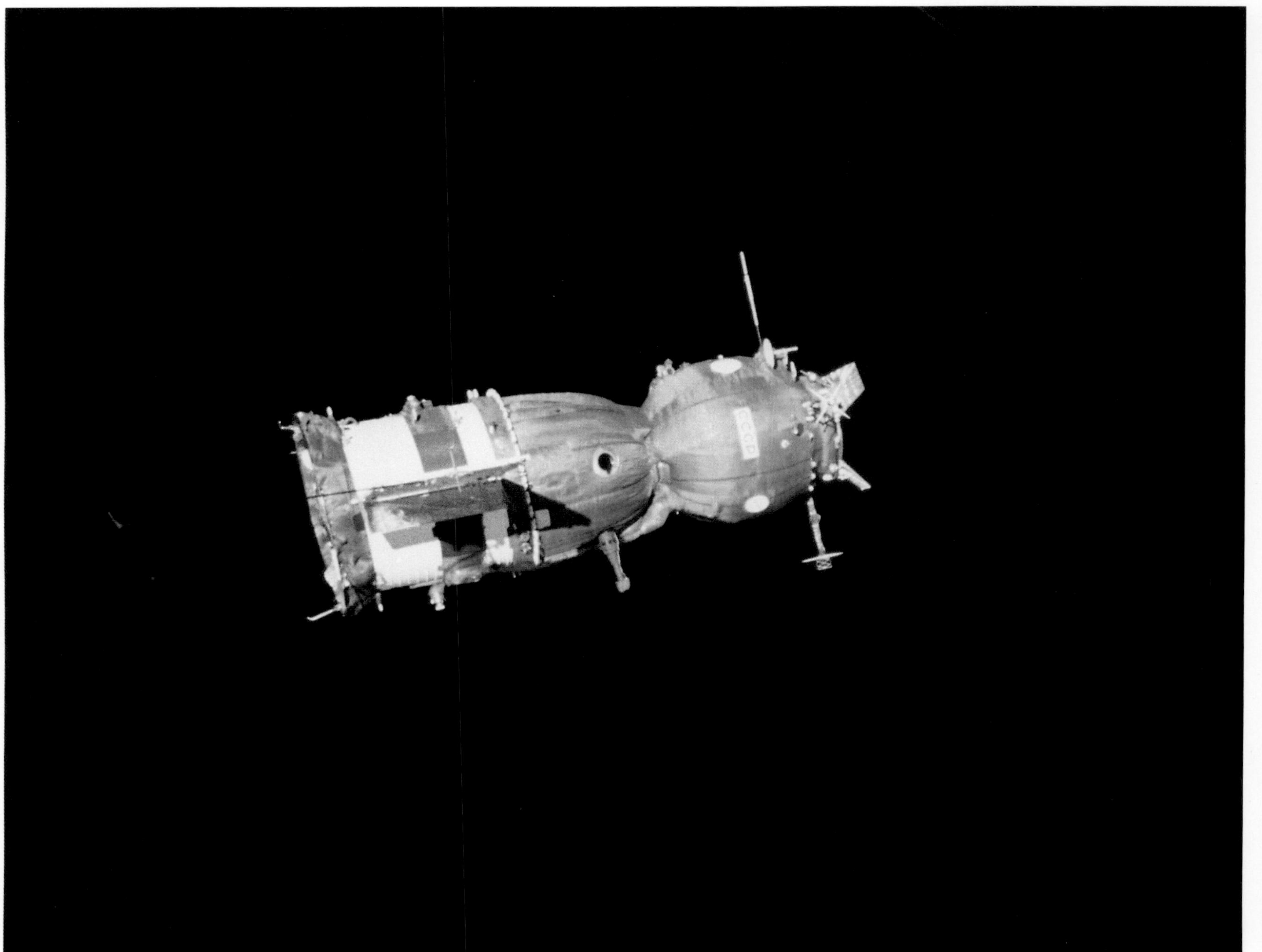

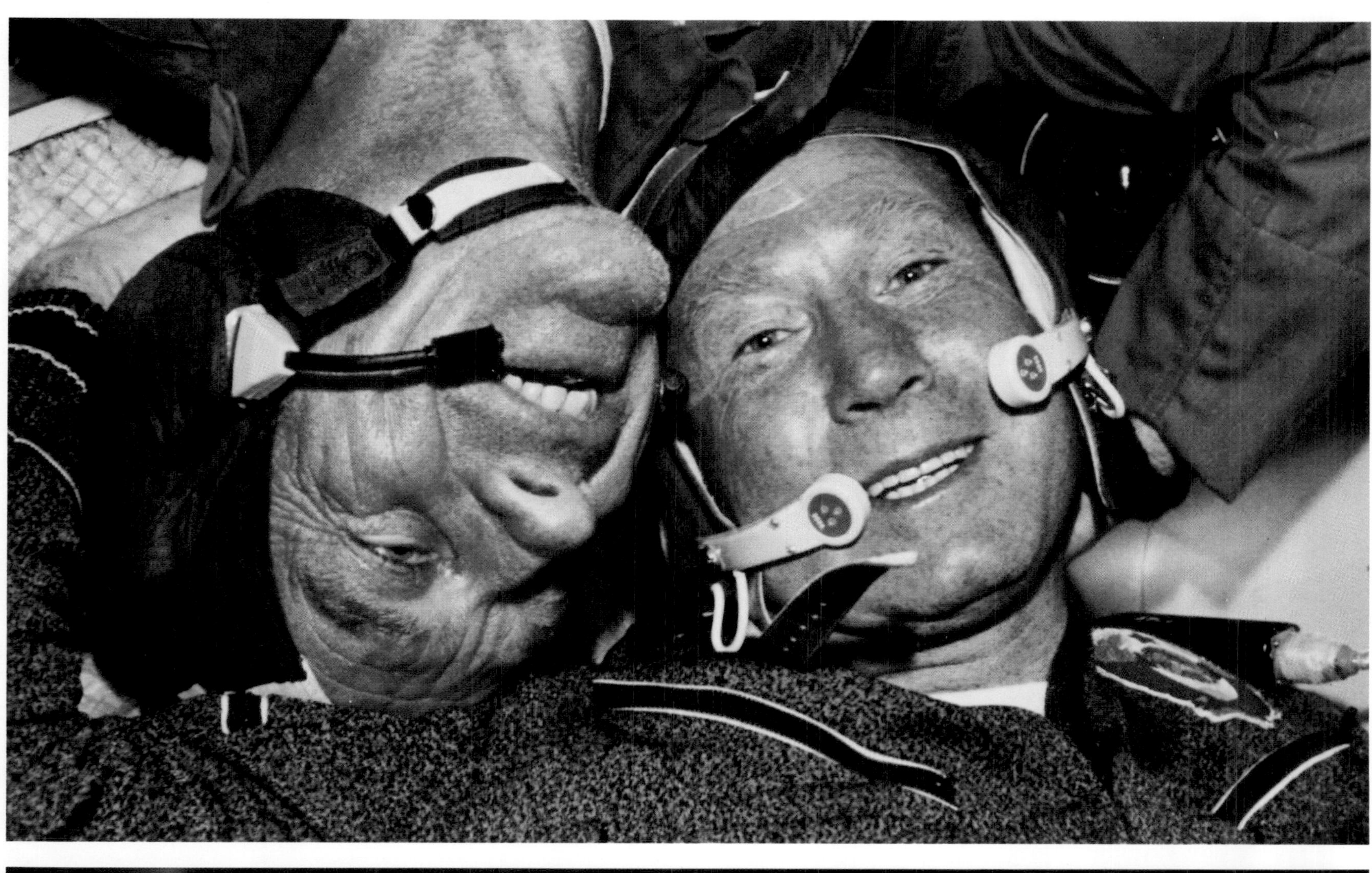

The Space Transportation System

With the completion of the Apollo-Soyuz flight, NASA's era of one-mission manned spacecraft closed. NASA knew that to gain maximum benefit from space, it would need spacecraft that like airliners could repeatedly take off and land on earth.

With the Space Transportation System (STS), NASA inaugurated a new epoch in transportation between earth and space. STS enables us to do economically many useful things in space that not too long ago were considered impractical. It opens space to reasonably healthy men and women of all nations. (In May 1983, a NASA Task Force for the Study of Private Citizens on the Shuttle recommended that space flights be opened to private citizens.)

The Space Shuttle vehicle itself consists of four parts: the airplane-like Orbiting vehicle (Orbiter), a huge external liquid fuel tank and two reusable solid rockets (SRB). There have been five Orbiters built. *Enterprise* (OV-101), *Columbia* (OV-102), *Challenger* (OV-99), *Discovery* (OV-103) and *Atlantis* (OV-104). A fifth Orbiter was ordered in 1985 to replace OV-99 which was destroyed on 28 January of that year.

The Orbiter provides as many as seven people with airline-like comfort as they live and work in space. The passengers and crew can dress in ordinary clothing rather than cumbersome space suits.

The Orbiter has a spacious cargo bay 60 feet long and 15 feet in diameter. The Orbiter is designed to carry as much as 65,000 pounds of cargo into space. The bay can accommodate unmanned applications or scientific spacecraft for launch from the bay or a fully equipped manned laboratory designed for use in space. It also accommodates Getaway Specials, the popular name for items flown in the STS Small Self-Contained Payload Program. The program gives any person or organization in the world an opportunity to purchase, at a comparatively moderate price, space for a chosen payload to be flown in space.

The manned laboratory, called Spacelab, was designed and built by the European Space Agency. Another Shuttle device built by another country is the Canadian Remote Manipulator System, an arm-like mechanism in the cargo bay operated from the Shuttle flight deck to deploy objects into orbit or retrieve them.

In a typical mission, the Orbiter's main engine and the solid rocket boosters ignite simultaneously, lifting the Shuttle from the launch pad. At a predetermined altitude, the spent boosters separate and parachute into the ocean where they are recovered. The external fuel tank is jettisoned just before the Orbiter reaches orbital velocity. The tank enters the atmosphere and breaks up over a remote ocean area. After completing its mission in space, the Orbiter uses its maneuvering rockets to slow down and re-enter the atmosphere. Once past re-entry and in the atmosphere, it behaves like an airplane. Gliding to earth it lands. Computer-driven controls backed up by crew skills enable the Orbiter to make virtually pinpoint landings.

The Orbiter's maximum altitude is about 1000 kilometers (600 miles). To send payloads to higher orbits it uses the Air Force Inertial Upper Stage (such as used for TDRS in STS-26) or for lighter payloads, the Payload Assist Module (such as used for SBS-3 and Anik C-3 in STS-5).

The first Orbiter, *Enterprise*, was employed for the crucial deadstick approach and landing tests at Edwards Air Force Base, California. Conducted in 1977, the tests involved the *Enterprise* being carried on the back of a 747 aircraft to an altitude of about 22,000 feet where it was released and guided to a landing. Edwards' extremely long and wide runways along with the flat dry lakebed surrounding them made it a suitable spot for such tests. Six of the first seven Shuttle missions into space also landed at Edwards; one landed at White Sands Missile Range, New Mexico.

The test program was cautious. Before being flown free, *Enterprise* was carried five times to altitudes of 25,000 feet to check performance, stability, control and safety of the two-aircraft combination. In three captive flights, two NASA astronauts aboard *Enterprise* helped verify the most favorable separation techniques, refine crew procedures and evaluate *Enterprise*'s systems. Finally, five free flights in which *Enterprise* was released and successfully landed from the carrier aircraft by the crew were accomplished. Crewmen chosen by NASA for the approach and landing tests were Fred W Haise Jr, Joe H Engle, Gordon Fullerton and Richard H Truly.

There are no plans to prepare *Enterprise* for space flight. It has been used in various public exhibitions such as a tour of United States and European

Above right: **A Space Shuttle fuselage undergoing manufacture at Rockwell International.** ***At right:*** **A Space Shuttle main engine, one of three per vehicle. The first powered Space Shuttle flight was slated for 1979, but didn't occur until April of 1981.** ***At far right:*** **The heat-shielding, tiled underside of an Orbiter.**

cities and the U.S. aerospace exhibit at the Paris Air Show in 1983. It will be used for fit checks at Vandenberg Air Force Base which is building Shuttle launch, landing and supporting facilities, and ultimately delivered to the Smithsonian Institution's National Air & Space Museum.

Development of the Shuttle meant advancement in new technological frontiers. Among the pacing items were the three hydrogen-fueled main engines of the Shuttle. As part of the Orbiter, they had to be built for repeated use. They had to be throttleable. Moreover, their specific impulse had to be much higher than any yet made. Specific impulse is the thrust gained from a pound of propellant in a second. It is comparable to the miles per gallon measure of a motorist.

Another pacing item was the array of more than 30,000 individual tiles that replaced the typical heat shields used in Mercury, Gemini and Apollo. The tiles were supposed to last from flight to flight not be burned away like the heat shield. They were supposed to be flexible enough to avoid cracking and to bond securely to the metal of the Orbiter. The tiles were made of a material that could be red hot on one side and cool enough to touch with one's bare hand on the other.

Perfecting these highly advanced items to assure success of the Shuttle from the first flight onward took time. By 1981, however, NASA was ready for the first manned Shuttle flight into space.

A new era in space, promising countless benefits for people everywhere, opened at 1:21 pm EST, 14 April 1981. At that time, the Orbiter *Columbia* crewed by astronauts John W Young and Robert L Crippen made a perfect landing at Edwards Air Force Base, California, after a nearly flawless space voyage. STS-1 was planned as a short flight in the interest of safety. Its major objectives were safe ascent, orbital flight and landing. Launched 12 April 1981, STS-1 was the first of four planned orbital-flight tests.

Although brief, STS-1 was a record-breaking flight. It was the first time an American manned spacecraft had been orbited without prior tests. It was the first launch of a true aerospace craft — a craft with wings and landing gear that would go into and return from orbit. It was the first time a spacecraft and its boosters (the two solid rockets that helped launch *Columbia*) were recovered for reuse. It was the first airplane-like landing of a craft from space.

While in space and during descent to earth, Young and Crippen found *Columbia* easy to control. They tested various systems, including the computers and opening and closing of the cargo bay doors. (Opening the doors is critical to deploy the radiators that keep heat from building up in the crew compartment. It is also critical to launch and retrieval of spacecraft and for conducting experiments. Closing them is vital to a safe return to earth.) The crew wore ordinary coveralls while in orbit.

The launch of *Columbia* again, for STS-2 on 12 November 1981, made it the first vehicle used for more than one space mission. In space, the crew — Joe H Engle and Richard H Truly — tested the Canadian-built Remote Manipulator System. This is the huge mechanical arm used to deploy spacecraft from the payload bay, retrieve spacecraft and do other freight-handling work in the bay. STS-2 also conducted experiments which demonstrated that it could be a stable platform for conducting earth surveys,

Below: The fourth in a series of captive test flights with prototype Orbiter _Enterprise_, which was an atmospheric test vehicle. After the five captive flights, the _Enterprise_ flew five atmospheric free flights, as is shown _at right_.

NASA
Enterprise
United States

NASA
905

tested advanced instruments and techniques for earth survey from space and gathered data about earth's resources and environment.

The STS-2 mission was cut short by failure of one of three fuel cells that convert hydrogen and oxygen into drinking water and electrical power. Mission safety rules for STS-2 required that if one of its fuel cells malfunctioned, the mission had to be ended. The crew landed *Columbia* at Edwards on 14 November 1981.

Preliminary inspection of *Columbia* indicated it came through STS-2 in even better condition than STS-1. No tiles were lost and only about a dozen were damaged and needed to be replaced. The improvement is attributed largely to a shock-absorbing water spray system installed on the launch pad and to improved bonding of tiles in the most vulnerable areas.

In its busiest and longest flight to date, *Columbia* was put through exacting tests and gathered a rich lode of useful space science, medical and materials-processing data. Launched 22 March 1982, Astronauts Jack R Lousma and C Gordon Fullerton worked in space for eight days. They landed *Columbia* at an alternate landing site — White Sands Missile Range, New Mexico — because heavy rains had drenched the dry lake bed at Edwards Air Force Base, the primary landing field.

While in space, they pointed different parts of *Columbia* to the sun for prolonged periods in thermal tests, repeatedly turned the orbital maneuvering engines on and off, and operated the Remote Manipulator System. They solved or adjusted to such comparatively minor problems as space adjustment syndrome (space sickness), a balky toilet, temperature control and radio static.

STS-4, the fourth *Columbia* mission, and the last orbital-flight test was marked by an on-time launch, nearly flawless completion of mission objectives and a perfect pinpoint landing on a concrete runway at Edward Air Force Base. Landing on the runway rather than the dry lakebed as before meant that the landing had to be precise.

Thomas K Mattingly and Henry W Hartsfield Jr were the crew. Their eight-day flight which began 27 June 1982, was likened by President Reagan to the 'golden spike' that opened transcontinental railroading. *Columbia*'s nearly flawless performance resulted in certification of the Space Transportation System as operational. On STS-4, *Columbia* carried the first commercial, Getaway Special and military experiments. In addition, Mattingly and Hartsfield participated in two medical experiments. The experiments were winning entries of the Shuttle Student Involvement Project of NASA and the National Science Teachers Association.

The Space Transportation System literally opened for business with the STS-5 flight of *Columbia*, 11 – 16 November 1982. This first operational flight also marked the first launch of satellites from *Columbia*, the commercial communications satellites SBS-3 and Canada's Anik C-3. This was also the first with a crew of four: Vance D Brand, commander; Robert V Overmyer, pilot; and Joseph P Allen and William P Lenoir, mission specialists. Mission specialists are qualified in satellite deployment, payload support, extravehicular activities (EVA) and the operation of the Remote Manipulator System. The latter was not planned for use on this flight, and mechanical failures in both space suits cancelled the planned space walks of Allen and Lenoir. In addition to its primary mission of launching the commercial communications satellites, STS-5 conducted three Shuttle Student Involvement Projects and one Getaway Special experiment.

Challenger's inaugural flight in the Space Transportation System was STS-6 in April 1983. *Challenger* had originally been a test article, and like *Enterprise*, not intended for space flight, but it was later reconfigured to be capable of missions in space.

Paul J Weitz was commander of STS-6; Karol J Bobko, pilot; and Story Musgrave and Donald K Peterson, mission specialists. Musgrave and Peterson carried out the first EVAs since Skylab in 1974. Moving about and working in *Challenger*'s open cargo bay, they practiced techniques for future missions.

STS-6 launched NASA's first Tracking and Data Relay Satellite (TDRS), which is part of a system that will vastly augment ground communication with earth-orbiting spacecraft and enable NASA to close most of its ground stations. The Inertial Upper Stage failed, however, to place TDRS in the required orbit. NASA ground controllers over a period of many weeks steadily nudged TDRS into the required orbit using surplus gas in the satellite's stabilization system.

STS-6 continued experiments in electrophoresis and in production of monodisperse latex microspheres which have important applications in the

At right and above right: **Space Shuttle *Columbia* is shown as attached to its two solid fuel booster and the huge, white-painted, oxygen-hydrogen fuel tank that saw use in the first two Shuttle flights, but was replaced with an unpainted (red primer) tank. *At far right:* STS-1, the first Shuttle launch, on 12 April 1981.**

United States
USA

pharmaceutical and medical fields. It also carried three Getaway Specials involving artificial snow crystals, packaged seeds and metals research.

The crew of the STS-7 (*Challenger*) mission of 18-24 June 1983, used the Remote Manipulator System to release a spacecraft called the Shuttle Pallet Satellite (SPAS) into orbit and to retrieve it. They practiced this five times before finally stowing SPAS in the cargo bay. While the satellite was in orbit, the astronauts flew *Challenger* around, above and below it and as far as 1000 feet away. During these maneuvers, cameras on both craft telecast spectacular pictures of each other against backdrops of black space and blue earth.

In addition to this important first, STS-7 featured two others: the first five-person crew and Dr Sally K Ride, first American woman in space. Other STS-7 crew members were Robert L Crippen, commander; Frederick H Hauck, pilot; and John M Fabian and Norman E Thagard, mission specialists.

STS-7 also launched two communications satellites: Canada's Anik C-2 and Indonesia's Palapa B. In addition, a number of experiments were carried out on *Challenger*. Among them were the OSTA-2 materials-processing experiment package of NASA and West Germany; a NASA Monodisperse Latex Reactor for producing microspheres that can be used in medical research and industry; a McDonnell-Douglas Continuous Flow Electrophoresis System for producing pharmaceuticals and medical research products by separating biological materials; and seven Getaway Specials covering such areas as crystal growth, soldering, germination of sunflower seeds and behavior of members of an ant colony.

STS-7 was a nearly flawless mission. Bad weather at Kennedy Space Center, Florida, prevented *Challenger* from making its planned landing there. Instead, *Challenger* was rerouted to Edwards Air Force Base.

Challenger made its third flight in a little over four months on 30 August 1983. Like STS-7, this flight lasted six days and had a crew of five. Also like its predecessor, STS-8, it was recognized for a handful of firsts. It was the first time that the Shuttle had been launched at night, and it carried the first black astronaut to make a space flight, Guion Bluford Jr. The rest of the crew included Daniel Brandenstein, Dale Gardner, William Thornton and Richard Truly, who was the second man to fly twice aboard the Shuttle.

Columbia flew mission STS-9, its first flight in just over a year. Launched on 28 November 1983, the record 10-day mission carried the European-designed Spacelab for the first time. For STS-9 *Columbia* was flown by Brewster Shaw Jr and John Young, the first American to make six space flights. Operating the Spacelab hardward were mission and payload specialists Owen Garriott (Skylab 3's scientist astronaut), Bryon Lichtenberg and Robert Parker, as well as ESA scientist Ulf Merbold, the first non-American astronaut to fly aboard an American spacecraft.

STS-9 was the last mission to receive an STS number and the first to carry a number under NASA's cumbersome new three-digit designation system. The first digit referred to the fiscal year (FY) in which the flight was scheduled, the second to the launch site and the third to the order in which the mission was originally scheduled within that fiscal year. Thus it came to be that STS-9 was also 41-A, with the 4 designating FY 1984 (NASA's fiscal years start in October) and the 1 designating Kennedy Space Center (2

During *Challenger* STS-7, in June of 1983, crewmembers used the Remote Manipulator Arm to launch the Shuttle Pallet Satellite SPAS, as is shown *below*. Another Shuttle satellite launch method is shown *at right*, wherein business satellite SBS-4 was rotated out of its cradle in the *Discovery's* payload bay in August of 1984.

HUGHES
SBS
NASA

is for Vandenberg AFB in California, which will not host a manned space flight until after 1990). The A indicated that it is the first flight scheduled in FY 1984.

The first flight of calendar year 1984 was the launch on 3 February of *Challenger*. Designated 41-B, the mission lasted eight days and carried a crew of five. During the mission, Bruce McCandless made the first EVA using the Manned Maneuvering Unit (MMU), a backpack-type device that permitted an astronaut to walk in space completely independent of the spacecraft. The sight of McCandless floating free high above the earth was inspiring.

Mission 41-C was launched on 6 April 1984, with the expressed mission of making repairs, in this case of the Solar Maximum Mission spacecraft. The Solar 'Max' had been launched in February 1980 to observe solar phenomenae during a period of maximum solar flare activity and had malfunctioned in December of that year. During Mission 41-C the satellite was retrieved by *Challenger*'s remote manipulator arm and successfully repaired by astronauts George Nelson and James van Hoften in the longest EVA since Apollo 17. Solar Max was then re-released into space to continue its observations of the sun and later Halley's comet.

Mission 41-D, launched on 30 August 1984 for a five-day flight, was the maiden voyage of the Orbiter *Discovery* (OV-103). The crew included America's second woman in space, Judy Resnik as well as Charles Walker, the first American to go into space representing private industry. Walker's job was to operate the McDonnell Douglas Electrophoresis Operations in Space (EOS) project. EOS had been developed by McDonnell Douglas as a means of processing pharmaceutical materials in the weightlessness of space.

A month after *Discovery* returned from 41-D, *Challenger* was launched on 41-G (missions 41-E and 41-F were cancelled). The mission was launched on 5 October 1984 and included the first Canadian astronaut, Marc Garneau. The mission was also Sally Ride's second into space, and it was the first flight by America's third woman in space, Kathryn Sullivan. Sullivan was also the first woman to walk in space, and Paul Scully-Power became the first oceanographer to scientifically observe the earth's oceans from space.

The last flight of 1984 and the first of FY 1985, 51-A, was launched on 8 November. For three months running, the United States had launched a manned space flight every month. Even after 14 flights, the STS seemed like it was chalking up new milestones on every flight and 51-A was no

Challenger 41-B astronaut Bruce McCandless II is shown *at right* during the first EVA to use the Manned Maneuvering Unit (MMU). *Below:* Space Shuttle *Columbia* touches down at Edwards AFB, demonstrating the facility with which the Shuttles return to earth. *At far right:* The historic night launch of *Challenger* STS-8.

different. During this mission, the crew located and retrieved the Palapa B-2 and Westar 6 satellites that had misfired into useless orbit during 41-B nine months before. The 'rescue' of the two wayward satellites made the commercial potential of the Shuttle seem much greater than ever before.

The year 1985 saw an unprecedented total of nine American manned space flights. The first was 51-C, *Discovery*'s three-day semiclassified Defense Department mission launched on 24 January. The second flight of the year, 51-D, was the launch of *Discovery* on 12 April, carrying the American space program's first VIP passenger, Senator Jake Garn (Republican, Utah) whose Senate committee oversees NASA's budget. Also on board were Charles Walker, from the McDonnell Douglas EOS project, who had first flown in space less than a year before, and Rhea Seddon, the fifth American woman to fly in space in less than two years.

Mission 51-B, which had first been scheduled earlier in the year, was launched on 29 April, less than two weeks after 51-D landed. Even as *Discovery* was flying the 51-D mission, the ESA Spacelab module was being loaded aboard *Challenger* for 51-B. The 51-B crew included Robert Overmlyer and Frederick Gregory as commander and pilot, respectively.

Mission 51-G was launched on 17 June 1985 with Daniel Brandenstein as commander, John Creighton as pilot, three NASA astronauts as mission specialists, two non-American astronauts as payload specialists, Patrick Baudry of CNES, the French space agency, and Sultan Salman Abdel-aziz Al-Saud of Saudi Arabia.

On 12 July 1985, *Challenger* was set to go on its eighth space flight and the fiftieth of the American manned space program, when a minor engine malfunction forced a delay. The launch of 51-F came 17 days later as *Challenger* carried the ESA Spacelab into orbit for the third time. Spacelab's solar telescopes provided the most detailed observations of the sun and other deep space objects since Skylab, 12 years before. For the first time, Spacelab flew without an ESA astronaut as payload specialist. The commander and pilot for 51-F were Charles G Fullerton and Roy Bridges Jr, both US Air Force colonels.

Mission 51-I, which saw the 26 August launch of the Orbiter *Discovery*, was the twentieth launch of the Shuttle program. The commander and pilot were Joe Engle and Richard Covey, and the astronauts included James van Hoften, Michael Lounge and William Fisher, whose wife Anna Fisher had flown aboard 51-A nine months before. During the eight-day mission, Fisher and van Hoften made two space walks to repair and redeploy the Leasat F3 satellite that had been launched during 51-D in April but had failed to activate.

Mission 51-J, launched on 3 October 1985, was the first flight of the Orbiter *Atlantis* (OV-104), the fourth of the space-rated Shuttle Orbiters. The mission was the second Defense Department all-military flight (51-C was the first), whose purpose was to launch a pair of Defense Satellite Communications System (DSCS) 'Discus-3' satellites. During the four-day maiden voyage, *Atlantis* and its crew set a Space Shuttle program altitude record of 320 miles, which was about 50 percent of the theoretical

Onboard scenes *Below:* The *Challenger* 41-B crew—clockwise from upper right: Dr Ronald E McNair, Robert Gibson, Vance Brand, Robert L Stewart and Bruce McCandless II. *At right:* Senator Jake Garn, aboard *Discovery* 51-D. *At far right:* Dr Sally K Ride, the first American woman in space, conducts a test as part of *Challenger* STS-7. *Below right*: Astronaut Anna Fisher prepares for sleep, aboard *Discovery* 51-A.

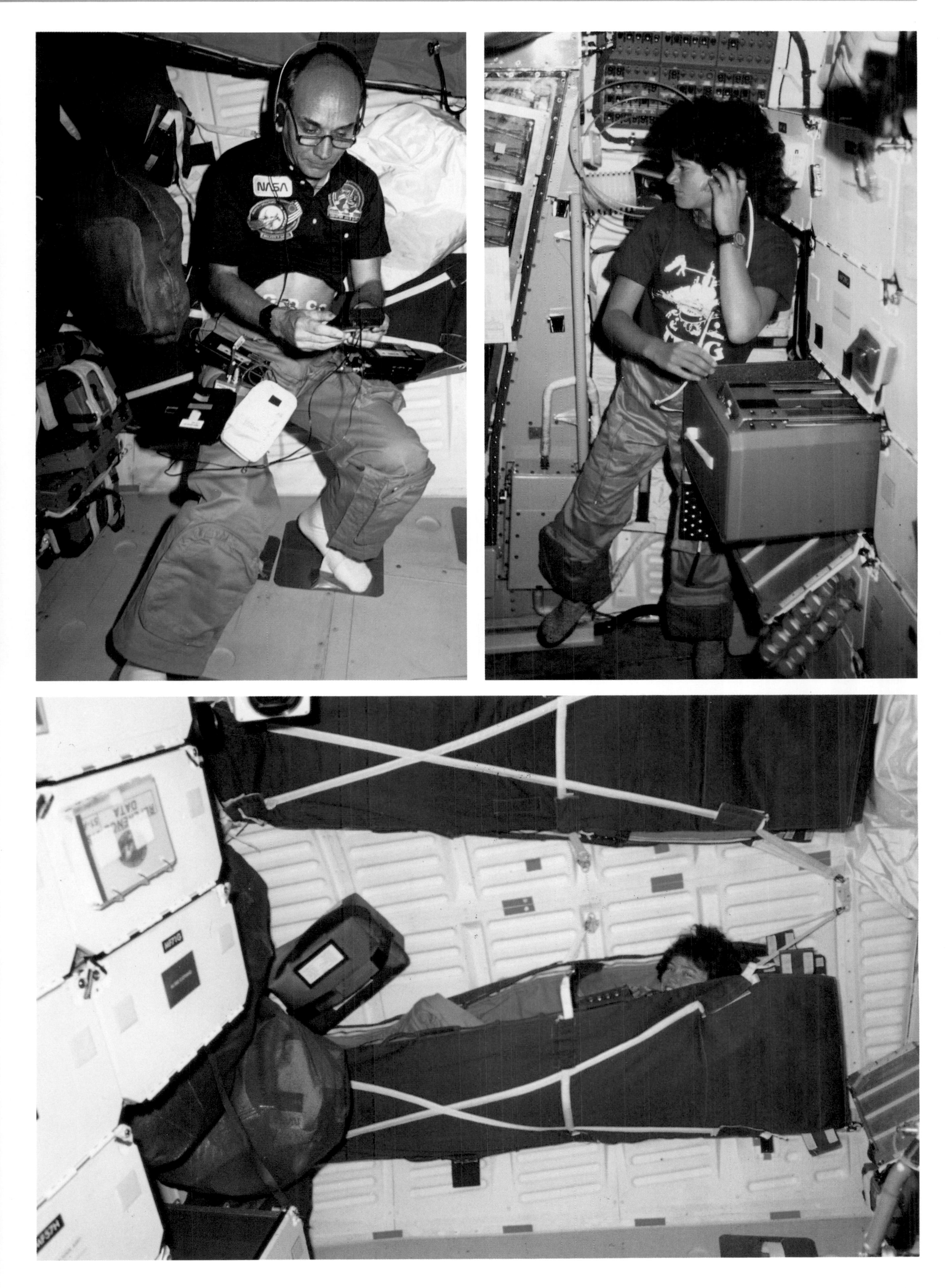
NASA

maximum altitude of a Shuttle Orbiter. In September 1987, *Atlantis* was planned to fly to an altitude of 368 miles for observations with NASA's Hubble Space Telescope during a mission that would see the first major observatory to observe deep space from space, but that mission would never take place.

Mission 61-A, launched on 30 October 1985, saw the ESA Spacelab flown for the fourth time, this time with the payload designation Spacelab D-1. It was to be *Challenger*'s ninth and last mission.

On 26 November 1985, *Atlantis* thundered into the sky in the spectacular nighttime launch of mission 61-B. The mission commander was Brewster Shaw Jr, with Bryan O'Conner as pilot. As astronaut Mary Cleave operated the remote manipulator arm from inside *Atlantis*, astronauts Jerry Ross and Sherwood Spring went out into the open payload bay, where they began assembling a 45-foot truss tower composed of 93 tubular 6½ and 4½ foot beams.

Nine American manned space flights took place in calendar year 1985, nearly double the number of 1965, 1966 and 1984, which were tied for second with five apiece. More Americans had flown into space aboard the Shuttle during 1985 than had flown in the entire Apollo program from 1968 through 1975. By the year end the four Orbiters had made 23 flights between them. *Challenger* (OV-99) had made nine, *Columbia* (OV-102) had made six, *Discovery* (OV-103) had made six and *Atlantis* (OV-104), which had first flown in October 1985, already had two space flights under its belt. During the year a number of non-NASA Americans went in orbit, either as technicians representing American aerospace firms or foreign governments, and the first US senator, Garn of Utah, made a flight into space.

On 12 January 1986, the Shuttle Orbiter *Columbia* was launched into space for the first time since November 1983. The first mission of the year, the seventh flight of *Columbia* had originally been scheduled for 18 December, but had to be postponed six times. Aboard *Columbia* for the six-day mission 61-C were Robert Gibson as commander, Charles Bolden Jr as pilot and three NASA astronauts as mission specialists, George Nelson, Steven Hawley and Franklin Chang-Diaz. RCA payload specialist Robert Cenker was aboard to help launch an RCA Ku-band communications satellite. Flying as an observer was US Representative Bill Nelson (Democrat, Florida), chairman of the House Science and Technology subcommittee on space science and applications.

In calendar year 1986 NASA hoped to see its new Space Flight Participant Program begin, under which private citizens not associated with a governmental agency or aerospace firm would be selected to make space flights. As requested by President Ronald Reagan, the first participant in such a program was an American classroom teacher. By the February 1985 deadline, 11,146 teachers had submitted their 11-page applications to the space agency. From this group, NASA selected Christa McAuliffe, a 37-year-old high school social studies teacher from Concord, New Hampshire. The backup candidate, who was to undergo the same training, was Barbara Morgan, a 34-year-old second-grade teacher from McCall, Idaho.

Christa McAuliffe's long-awaited flight was the Shuttle program's mission 51-L. Scheduled for 26 January 1986, 51-L would be the program's

***Below:* Atlantis 61-B astronauts Jerry Ross and Sherwood Spring, with one of the structures they built during the EASE tests. Another of the EASE structures is shown in construction *at right*.**

twenty-fifth flight and the tenth for the Orbiter *Challenger*. The mission commander was Francis 'Dick' Scobee, who had first gone into space aboard *Challenger* in April 1984. *Challenger*'s pilot for the 51-L flight was Mike Smith, a former Navy fighter pilot. The three NASA payload specialists were rookie Ronald McNair and veterans Ellison Onizuka and Judy Resnik, the second American woman in space. The non-NASA personnel included McAuliffe and Greg Jarvis, a former Air Force pilot representing Hughes Aircraft. The crew of 51-L represented a true cross-section of American life. Geographically, their home states spanned the nation from Hawaii to New Hampshire. There were two women, a black man and a Japanese-American. Both Scobee and Smith were former combat pilots, and Resnik was a classical pianist.

The center of attention, however, was Christa McAuliffe. Other American private citizens had gone into space before, but they had been either politicians or (like Jarvis aboard this flight) aerospace engineers. There was something about a high school teacher going into space that captured the imaginations of Americans of all ages. A special NASA video network was set up so that students in her own school back in Concord, New Hampshire, as well as thousands of others across the nation, could watch America's first space teacher go into space and teach her classes from earth orbit. It was to have been the first time that a 'regular' person would be speaking to the nation from space.

An icy cold front moved into Florida over the final weekend of the 51-L countdown, forcing the Sunday launch to be postponed to Tuesday, 28 January. Thus it came to be that students throughout the nation could watch the launch from their classrooms. Christa McAuliffe's own children—Scott, 9, and Caroline, 6—were on hand at Cape Canaveral with their father, Steve McAuliffe, and members of Scott's third-grade class, to watch *Challenger* go into space.

It was cold, just above freezing, as the seven members of the 51-L crew boarded the bus to take them out to the Orbiter on Pad 39B. As the crew entered *Challenger* they were smiling and appeared relaxed. A member of the ground crew handed teacher McAuliffe a shiny red apple as they climbed aboard. The textbook-perfect launch came at 11:38 am. *Challenger* cleared the launch pad and climbed slowly into the sky. All the Shuttle's engines seemed to be running smoothly and Smith was in touch with ground control as he throttled down, right on cue. In Concord, New Hampshire, Christa McAuliffe's class cheered as they watched *Challenger* on television.

Challenger was at an altitude of 47,000 feet and traveling at a speed of 1800 feet per second when a small flame appeared at the base of the right solid rocket booster, and began to lick at the enormous fuel tank with its half-million gallons of liquid hydrogen and liquid oxygen. Fifteen seconds later and 73 seconds after launch, the Orbiter was enveloped in a fireball a hundred yards across as the fuel tank exploded. Both *Challenger* and the tank disintegrated in the explosion while the two solid rocket boosters twisted and turned across the sky like blinded, panicked animals. Across the nation, from Cape Canaveral, where they watched the explosion in person to Concord, where the scene was transmitted on live television, spectators stared dumbfounded, unsure of what had happened. Then, from mission control, came the confirmation of what everyone feared but hoped wasn't true:

> 'Flight controllers are looking very carefully at the situation. Obviously a major malfunction. We have no downlink (communication with *Challenger*) . . . The vehicle has exploded.'

Nine years and one day after the Apollo 1 fire, seven people died in what would have been NASA's fifty-sixth manned space flight, but which became instead the worst disaster in a quarter century of manned space flight.

For nearly an hour debris drifted down from the cloud that had swallowed *Challenger*. Paramedics and rescue teams were on hand before the last fragments plunged into the Atlantic, but there was no hope of survivors. Over the days that followed, NASA began to sort out what had happened. The 14 Shuttle missions that were to have followed 51-L during 1986 were put on hold as the explosion was investigated and steps were taken to redesign the Solid Rocket Boosters and revise NASA launch procedures. Ultimately, the Shuttle Transportation System that had promised so much in the early 1980s, was grounded for over two and a half years.

At above right, an anomalous flame appears above the exhaust of the _Challenger_ 51-L Solid Rocket Boosters; _above far right_, the fire spreads and, as _Challenger_ 51-L Mission Control _(below)_ helplessly watches, _Challenger_ explodes _(below right)_, claiming the lives of its crewmembers—who are named in the text on this page.

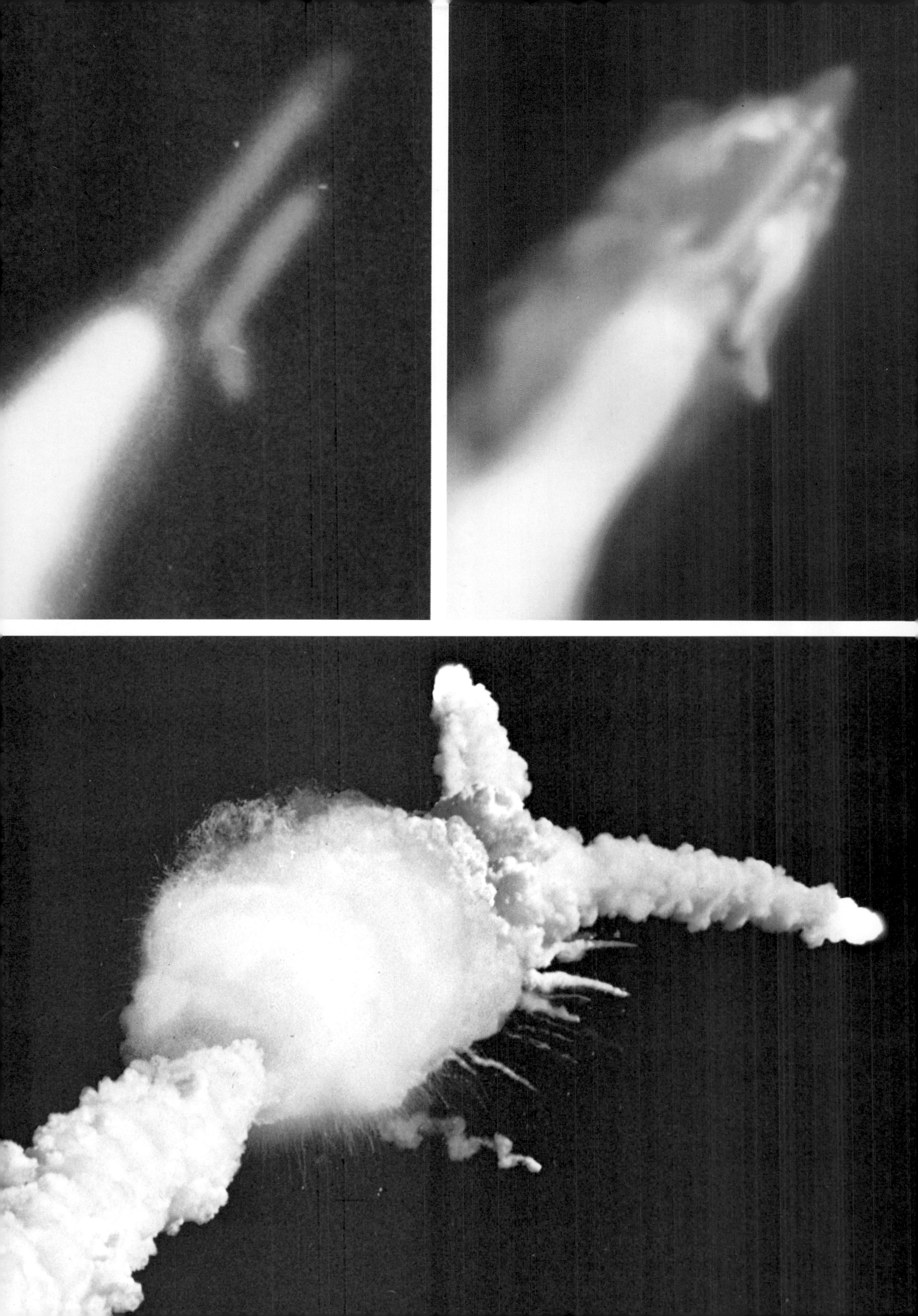

1977 Flight Tests STS-1 (1981) STS-2 (1981) STS-3 (1982) STS-4 (1982) STS-5 (1981)

41-C (1984) 41-D (1984) 41-G (1984) 51-A (1984) 51-C (1985)

The Space Shuttle Transportation System 1977–1986

In 1977, the US began its successful atmospheric glide tests of the Shuttle Orbiter prototype *Enterprise*, but the *Space* Shuttle project suffered delays—a Shuttle Orbiter was to enter orbit sometime in 1979, but did not do so until, on 12 April 1981, Astronauts John Young and Robert Crippen took off in the Shuttle Orbiter *Columbia* and guided their craft through a 54-hour orbital mission to a perfect ground landing.

***Columbia* returned to space on 12 November, and flew three times in 1982. In 1983, Sally Ride became the first American woman in space, Guion Bluford became the first black American man in space and West Germany's Ulf Merbold became the first non-American to fly aboard an American spacecraft. The latter mission also saw the first use of the ESA Spacelab Module. The first three missions of 1983 were flown aboard the second Shuttle Orbiter, the freshly inaugurated *Challenger*.**

***Challenger* was followed by the inaugurations of two more Shuttle Orbiters—*Discovery* in 1984 and *Atlantis* in 1985. The reusable, and heavily used, Shuttle Orbiters were employed for scientific tests; launched, rescued and repaired satellites, and set the record for manned space flights in a single year at nine in 1985. The year 1986, as is discussed in the following chapter, brought tragedy to the Shuttle Orbiter program.**

After the *Challenger*/ 51-L tragedy, several years were consumed in testing and re-testing the STS equipment and systems for safety. At this point, general plans are in the works for a Shuttle mission sometime near the end of the present decade.

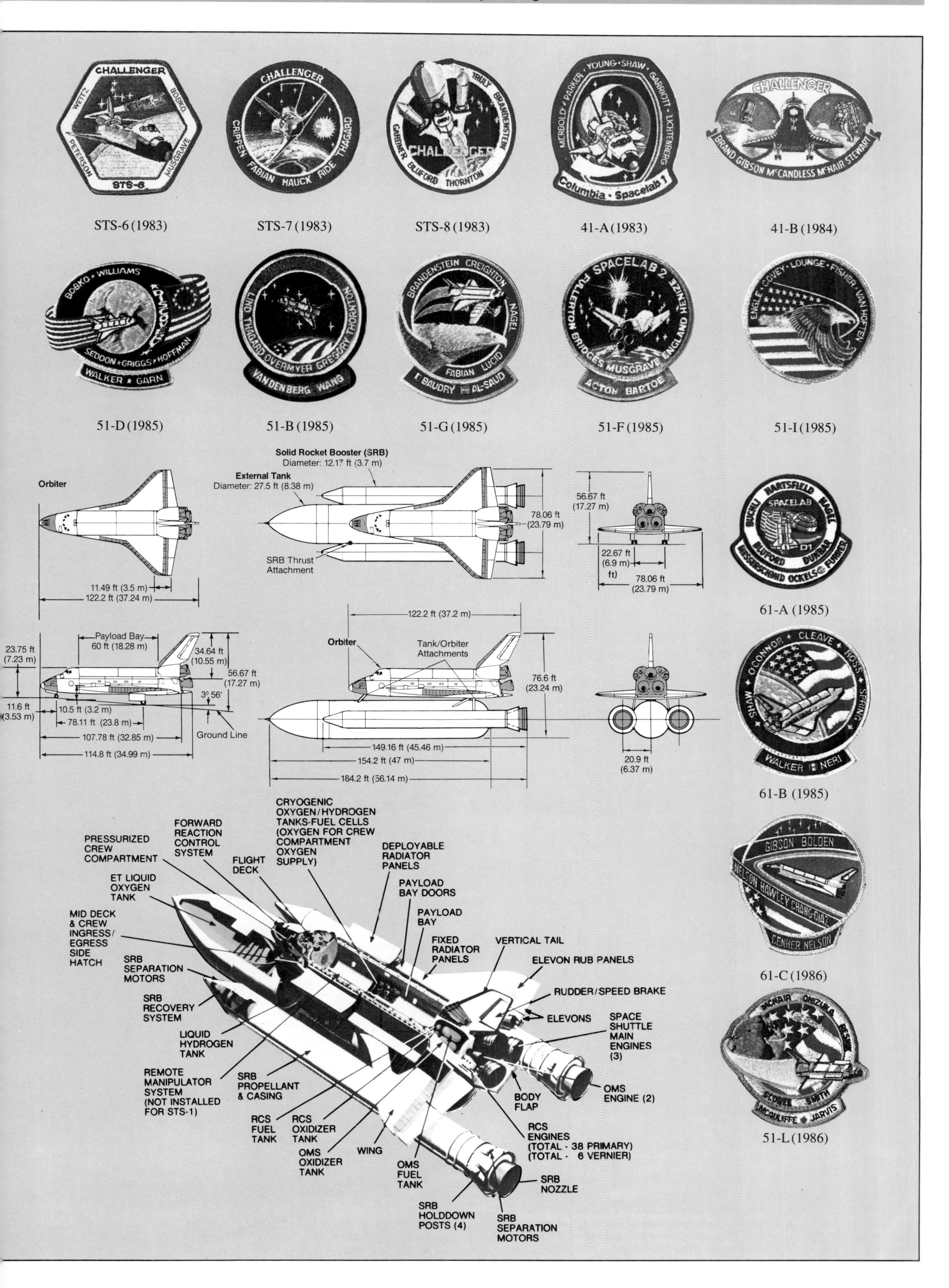

STS-6 (1983) STS-7 (1983) STS-8 (1983) 41-A (1983) 41-B (1984)

51-D (1985) 51-B (1985) 51-G (1985) 51-F (1985) 51-I (1985)

61-A (1985)

61-B (1985)

61-C (1986)

51-L (1986)

NASA TODAY

Edited by Bill Yenne

With manned spaceflight as its most visible activity, NASA's image was dealt two serious blows by the 1986 Mission 51-L disaster which had destroyed the Spaceshuttle Orbiter *Challenger*, and by the subsequent 30-month suspension of *all* American manned space missions which seriously damaged NASA's public image as well as its internal morale. Resumption of flights in 1988 will have been a major boost in restoring a healthy glow to the perception that is held of America's space agency.

Recovery Begins

The investigation of 51-L pointed up not only serious flaws of the Shuttle System Solid Rocket Boosters, but serious flaws in NASA's procedures. All of these problems would have to be corrected before the Space Shuttle Transportation System would ever fly again.

However, while the agency concentrated on the resumption of Shuttle flights which NASA then hoped would happen in early 1988, major accomplishments in other areas were taking place. Following a spectacular encounter with Planet Uranus in January, the Voyager 2 spacecraft continued its scientific journey through the solar system toward an encounter with Neptune in 1989. Meanwhile, a new baseline configuration for the space station was adopted and draft requests for proposals were issued to prospective contractors — a major milestone for beginning development in 1987.

In 1986, NASA and the Department of Defense initiated the joint National Aero-Space Plane research program. This program will lead to an entirely new family of aerospace vehicles capable of horizontal takeoff and landing, single-stage operations to orbital speeds and sustained hypersonic cruise within the atmosphere using airbreathing propulsion.

Dr James C Fletcher became administrator of NASA for the second time on 12 May 1986, succeeding James M Beggs who had resigned. Fletcher had previously served as NASA administrator from April 1971 to May 1977. The agency was deeply involved in the investigation of the 51-L/*Challenger* accident, when Fletcher assumed office, and his statement upon receiving the report of this commission set the tone for his administration:

'Where management is weak, we will strengthen it,' he said. 'Where engineering or design processes need improving, we will improve it. Where our internal communications are poor, we will see that they get better.'

Fletcher immediately sought the opinions and advice of a large number of persons both in the agency and out, then initiated a major study of NASA management which has led to fundamental restructuring of two major programs and is expected to lead to important changes in the overall agency management structure.

Working with the National Academy of Public Administration, Fletcher appointed retired US Air Force General Samuel C Phillips, who had headed the Apollo program in 1964, to direct this study. In addition, three separate committees of the National Research Council were organized to provide oversight of Space Shuttle redesign efforts.

In other important activities during the year, Fletcher secured Presidential and Congressional support for a fifth Orbiter to replace *Challenger*, and he secured Presidential and Congressional support to keep space station development on track toward its original goal to achieve a permanent manned presence in space in 1994. He initiated a completed reassessment of

Above right: **Johnson Space Center memorial services for the crew of *Challenger* 51-L. First American woman in space Dr Sally Ride was a member of the President's Commission on the Space Shuttle Challenger Accident, a delegation of which is shown *at right. At far right:* A Solid Rocket Booster test following the accident.**

space station design and assembly procedures which has lead to major changes in this program. He has encouraged continuing negotiations over space station issues with international partners. He also made a substantial number of changes among the agency's top management personnel, especially in the Space Shuttle and space station programs.

In one of the most important moves during 1986, Fletcher created the new Office of Safety, Reliability, Maintainability and Quality Assurance, also in response to a recommendation by the Presidential commission, to re-emphasize these areas in the wake of the *Challenger* 51-L accident.

The administrator established a group to determine NASA's response to the long-term goals for the US space program recommended by the President's National Commission on Space, and initiated an effort to develop short-term goals to take the agency to 1995. The latter effort is under the direction of Dr Sally Ride and was completed in August 1987.

On 19 June 1986 NASA Administrator Fletcher announced the decision to terminate the development of the Centaur upper stage for use aboard the Shuttle. This decision was based on the fact that, even following certain modifications identified by ongoing reviews, the resultant stage would not meet safety criteria being applied to other cargo or elements of the Space Shuttle system.

On 15 August President Reagan announced his decision to support a replacement for *Challenger*. At the same time, it was announced that NASA no longer would launch commercial satellites, except for those which are Shuttle-unique or have national security or foreign policy implications.

On 22 August NASA announced the beginning of a series of tests designed to verify the ignition pressure dynamics of the Space Shuttle solid rocket motor field joint. The series was conducted over the course of next year at Morton Thiokol's Wasatch Division in Utah and Marshall Space Flight Center.

On 3 October 1986, NASA announced February 1988 as the target date for resuming Shuttle flights. It was a premature announcement, for the date would ultimately slip by over half a year. A 3-year projected manifest was also released, based on a reduced flight rate and accommodating as far as possible the payload backlog.

Toward Other Horizons

Nevertheless, 1986 was a year of progress and transition for the space station project as NASA laid the organizational and programmatic framework for beginning development during which the final design, construction, launch and initial operation of the permanently manned space station will take place. A new baseline configuration for the space station, called the 'dual keel,' was adopted as the reference configuration to guide the final 8 months of preliminary design activities. Former Lewis Research Center Director Andrew J Stofan was named Associate Administrator for space station in June and Dr Franklin D Martin was subsequently named the deputy associate administrator.

Ignition pressure dynamics tests *(below)* on the Solid Rocket Boosters were conducted at Morton Thiokol's Wasatch Division test range in Utah. *At right* is the Orbiter *Atlantis*, on a rollout for systems checking in October 1986. The Space Shuttle program would not, however, resume flight until September 1988.

USA

NASA later modified the station baseline configuration including expanding the 'resource' nodes used to connect the working and living modules together and established a revised assembly scenario. Expanding the resource nodes permits flight critical command and control equipment, previously located outside on the space station's framework, to be housed inside the nodes. This alleviates the need for crew members to perform EVA for routine maintenance and replacement of these components. The expanded nodes also provide about 4000 cubic feet of additional pressurized volume to the space station.

The temporary loss of NASA space launch capability precluded what was to have been 'A Year For Space Science.' Five major scientific mission launches were planned for 1986, including Spartan Halley, Astro-1 and three planetary missions—Galileo, Ulysses and the Hubble Space Telescope. However, NASA science and applications continued working a variety of activities not requiring launches.

In the area of Solar System/Planetary Science, the most notable achievement during 1986 was the successful encounter with planet Uranus by the Voyager 2 spacecraft in January. The Uranus encounter provided prime scientific data on a planetary body never before examined by a space probe at such close range. The 9-year-old robotic spacecraft, Voyager 2, is continuing its scientific journey through the solar system towards an encounter with planet Neptune in 1989.

An upgrade of the Deep Space Network (DSN) was key to the success of the Voyager 2 encounter with the planet Uranus in late January 1986. Nearly 500 images of the planet, its satellites and rings were obtained during the near-encounter phase. This support was possible due to a new method of arraying NASA's large antennas and by combining signals with Australia's large antenna at their Parkes facility. Also during March and April, the DSN completed its rather extensive support to the various Halley's Comet observations.

Work also began in 1986 to increase the sensitivity of the DSN in preparation for the Voyager 2 encounter with the planet Neptune in 1989. This will be accomplished by improving the efficiency of the large DSN antennas and by simultaneously combining signals received by the DSN antennas, during the encounter, with other antennas at non-NASA facilities.

The Galileo mission to Jupiter, a joint project with the Federal Republic of Germany, was planned to make a comprehensive, long-term study of the planet's atmosphere, magnetic field and its moons. The Galileo could be launched from the Shuttle either in November 1989, or June 1991, or be launched by an expendable launch vehicle.

The Ulysses mission, a cooperative effort between NASA and The European Space Agency, will provide the first view of the sun and the solar system from above the ecliptic plane. The data will provide knowledge about the sun and also will help scientists to better understand the effects of solar activity on the earth's weather and climate. The Ulysses mission was considered for launch in September 1989 or October 1990.

The Hubble Space Telescope originally scheduled for launch in October 1986, will carry five scientific instruments to study the stars, planets and intersteller space. Four telescopes are provided by the United States and the fifth by the European Space Agency. During 'down time,' the space telescope has undergone continual 'end-to-end' testing to maintain the health of the instruments.

During 1986, NASA's aeronautical research and technology efforts continued to expand American capabilities in civil and military aviation, contributing significantly to US world aviation leadership and to national security. These efforts covered the spectrum from fundamental disciplinary research to flight testing.

In President Reagan's State of the Union address, he said 'We are going forward with research on an aerospace plane that could shrink travel times between Washington DC and Tokyo, or any other cities no matter how distant. . . to less than 2 hours.' At the President's request, NASA and the Department of Defense initiated the joint National Aero-Space Plane (NASP) research program that will lead to an entirely new family of aerospace vehicles.

NASP is an accelerated technology development program leading to a flight research vehicle (X-30) to validate a wide range of aerospace technologies and capabilities including horizontal takeoff and landing, single-stage operation to orbital speeds and sustained hypersonic cruise within the atmosphere using airbreathing propulsion.

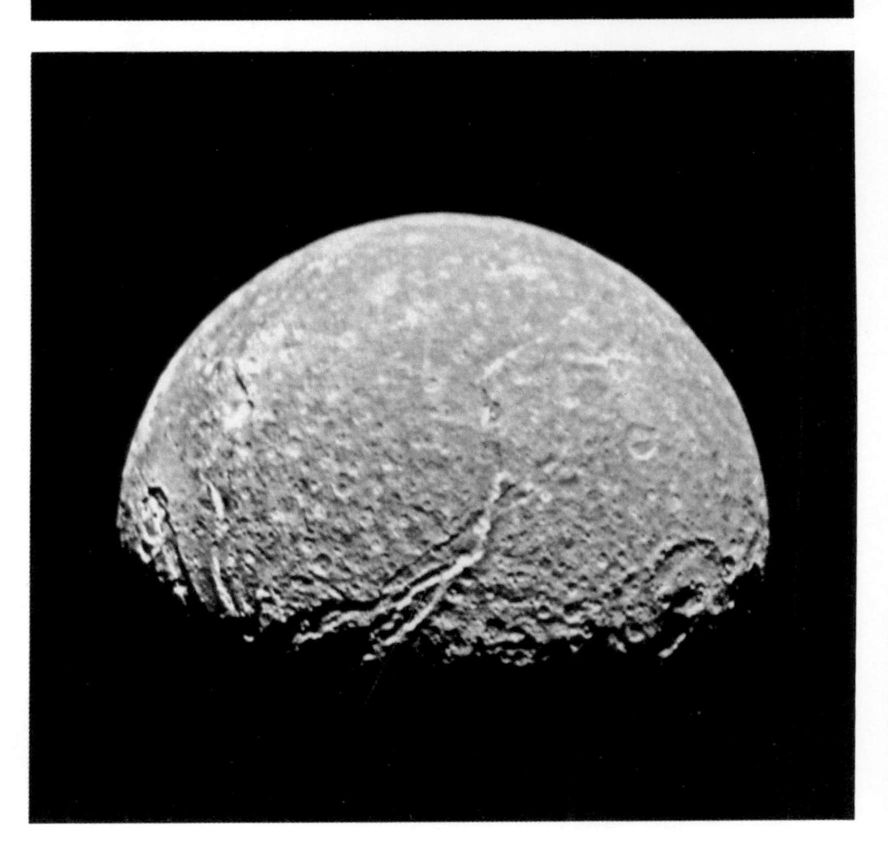

At right: **A 1986 Voyager 2 photograph of Uranus moon Titania. The rings of Uranus are clearly seen in the Voyager 2 photograph *at middle right. At top right,* A Voyager 2 departure photograph of Uranus' blue crescent. The Galileo Jupiter probe *(at above far right)*, scheduled for Shuttle launch in 1986, was postponed for three years. *At below far right* is an artist's conception of the Galileo launch.**

Galileo
NASA

Other joint NASA/DOD programs, such as the X-29 forward swept wing experimental aircraft, X-wing research aircraft, the tilt rotor/JVX aircraft and the mission adaptive wing, substantially augmented the military data base. Joint NASA/Federal Aviation Administration programs addressed lightning strikes, wind shear, icing and other issues affecting aviation safety. The X-29 aircraft completed its flight envelope at the Ames-Dryden Flight Research Facility, Edwards, California, in November. The X-29 is a joint Defense Advanced Research Projects Agency (DARPA)/US Air Force/ NASA flight research program.

In September 1986, only 6 days after launch and less than 24 hours after being put into operation, the Search and Rescue Satellite Aided Tracking equipment (SARSAT) on board the NOAA-10 satellite, picked up the first distress signal from a downed aircraft leading to the rescue of four Canadians who crashed in a remote area of Ontario. The RCA-built spacecraft was launched 17 September from Vandenberg Air Force Base, California, by a General Dynamics/US Air Force Atlas E launch vehicle. The Canadian/French-built search and rescue equipment on board the NOAA-10 satellite was activated on September 22, according to officials at NASA's Goddard Space Flight Center, Greenbelt, Maryland, where the SARSAT program is managed. The SARSAT equipment permits the satellite to pick up distress signals from aircraft or ships and to relay these signals to ground processing facilities which then dispatch rescue forces.

Beyond the results of Voyager at Uranus in January and isolated highlights such as NOAA-10, the year 1986 was NASA's most disappointing ever. Hopes of major spacecraft launches such as Galileo and the Hubble Space Telescope evaporated in the wake of the 51-L disaster and the loss of *Challenger*.

At right: **An artist's conception of the Galileo planetary probe entering the Jovian atmosphere. The Hubble Space Telescope, shown *at far right,* was another Space Shuttle launch that was scheduled for 1986, only to be postponed. *Below:* The Search and Rescue Satellite Aided Tracking system — see the text on this page.**

The Edwin P Hubble Space Telescope

Lockheed Missiles & Space

Placing a telescope in space has been a dream of astronomers for decades, dating back to before spaceflight was a fact. There, far above the distorting effects of the earth's atmosphere, astronomers would have an unimpaired view of the entire universe. The National Aeronautics and Space Administration is meeting this dream of astronomers through the Edwin P Hubble Space Telescope, a national observatory orbiting 320 nautical miles above the earth.

The Hubble Telescope is a 94.5 inch Ritchey-Chretien telescope, named in honor of American astronomer Edwin P Hubble, who made vital contributions to the understanding of galaxies and the universe through his work earlier this century.

While the mirror size and optical quality make the Hubble Telescope one of the largest and most precise astronomical instruments ever produced, its major advantage lies in the fact that it will be outside the earth's atmosphere.

The telescope's high vantage point will allow it to see farther and with greater clarity than any astronomical instrument ever built. Besides 'seeing' better it will detect more portions of the electromagnetic spectrum, such as ultraviolet light which is absorbed by the atmosphere before reaching the ground.

The heart of the Space Telescope is the Optical Telescope Assembly. The major segments of the OTA are the 94.5 inch primary mirror, a secondary mirror of 12 inch, and the OTA's support structure.

The precision of the primary and secondary mirror is a major ingredient in the superb capability of the Space Telescope. If the mirror were scaled up to the size of the earth, none of the great mountain ranges would tower more than five inches above the lowest point. Light entering the Space Telescope is reflected off the primary mirror to the secondary mirror, 16 feet away. The secondary mirror sends the light through a hole in the center of the large mirror, back to the scientific instruments.

Providing all the essential systems to keep the Hubble Telescope operating in the hostile environment of space is the function of the Support Systems Module. The SSM also directs communications, commands, power, and fine pointing control for the telescope. Collectively, the SSM consists of the light shield on the front end of the telescope; the equipment section, with the main spacecraft electronics equipment; and the aft shroud, which contains the scientific instruments.

The Fine Guidance Sensors feed roll, pitch and yaw information to the telescope's attitude control system. The pointing capability of the Hubble Telescope provided by the FGS is so precise it is often called a sixth scientific instrument. To point the telescope, the FGS must identify the position of specified stars. This pointing data can be used to calibrate space-distance relationships throughout the universe. The FGS will allow the telescope to point with a stability of 0.007 arc seconds, or roughly the equivalent of focusing on a dime in Los Angeles from a vantage point in San Francisco.

The Hubble Space Telescope carries five scientific instruments which are replaceable and serviceable in orbit. Four of the instruments, about the size of telephone booths, are located in the aft shroud, behind the primary mirror. They receive light directly from the secondary mirror. The fifth instrument, the Wide Field/Planetary Camera, is located on the circumference of the telescope and uses a pick-off mirror system.

The FOC does exactly what its name implies, observes faint objects. It does this by taking very low light levels and electronically intensifying the images. Objects as faint as 28th or 29th magnitude (the higher the magnitude, the fainter the object) should be observed by the FOC. By comparison, earth-based telescopes can see to about 24th magnitude. Likely targets for the instrument are the search for extra- solar planets, variable brightness stars, and in its spectrographic mode, the center of galaxies suspected of concealing black holes.

The FOS will measure the chemical composition of very faint objects. Visible light contains information used to determine the chemical elements which make up the light source. Special gratings and filters allow the FOS to make spectral exposures which not only reveal information about the makeup of a light source but also about its temperature, motion and physical characteristics. This instrument will study the spectra of objects in the ultraviolet and visible wavelengths. Particular targets of interest are quasars, comets and galaxies.

While performing in much the same way as the FOS, this instrument will observe only the ultraviolet portion of the spectrum. Ultraviolet light is filtered by the earth's atmosphere and the only spectrographic measurements taken in this light have been from previous space observatories. The HRS will be used to investigate the physical make-up of exploding galaxies, interstellar gas clouds, and matter escaping from stars.

With no moving parts, the HSP is the simplest of the five Space Telescope instruments. It will allow astronomers to take very exact measurements of the intensity of light coming from stellar objects. In addition, it will provide very precise measurements, down to the microsecond level, of time variations in the light. The amount of light received from an object is an important factor in determining its distance making the HSP useful in refining the scale of the Milky Way galaxy and other nearby galaxies.

Actually two separate cameras in one housing, the WF/PC should return some of the most spectacular visual images from the Hubble Space Telescope. In the Wide Field mode the instrument will view large areas of space and provide exquisite views of galaxies and star fields. In the Planetary mode, it will provide glimpses of the planets comparable to those obtained on close fly-by missions.

The Hubble Space Telescope will be placed in orbit by the Space Shuttle in the summer of 1989. That will be 320 nautical miles high and inclined 28.5 degrees to the equator.

After being deployed from the Shuttle cargo bay, the solar arrays and high-gain antennae will be deployed. The Shuttle will station-keep nearby while final checks are completed, and after a period of calibration and verification, astronomers at the Space Telescope Science Institute will begin their observations of the sky.

Provisions have been made for in orbit repair and maintenance of the Space Telescope. Astronauts can remove old instruments and replace them with more advanced or different types of devices. Repairs to many of the systems or instruments can also be accomplished in orbit. The Telescope can be returned to earth for major refurbishment and alteration if necessary. The operating life of the Hubble Space Telescope may be 15 years or more with refurbishment and modernizing using the Space Shuttle.

The Office of Space Science and Application at NASA Headquarters is responsible for overall program management, financial and scheduling provisions, and the science policy development and direction. Marshall Space Flight Center in Huntsville, Alabama is responsible for the development and operation of the Space Telescope system as the 'lead' NASA Center. In Greenbelt, Maryland, the Goddard Space Flight Center is managing the scientific instruments, mission operations, and data management. It is also charged with monitoring the Space Telescope Science Institute.

The Space Telescope Science Institute is operated by AURA, the Association of Universities for Research in Astronomy. The Institute is located on the Homewood Campus of Johns Hopkins University in Baltimore, Maryland. It is the job of the Institute to determine the observational program of the Space Telescope while in orbit, insuring that the observatory will be used to its maximum advantage.

Lockheed Missiles & Space Company, Sunnyvale, California, is the systems integrator of the Space Telescope satellite. It is also responsible for the design, development and manufacture of the Support Systems Module. Perkin-Elmer Corporation, Danbury, Connecticut, manufactured the Optical Telescope Assembly.

Below: **An artist's conception of the Hubble Space Telescope in orbit.**

The Cautious Optimism of 1987

The following year, 1987, however, was seen in more optimistic light. At least there were no disasters. 'NASA is on the road to recovery and eagerly anticipating the resumption of Space Shuttle flights in 1988,' said NASA Administrator Fletcher in his year end review. 'We made substantial advances in many areas last year,' Dr Fletcher said, 'but much of our work was actually building a foundation for 1988 and beyond. The United States can maintain a leadership role in space — with its tremendous importance to science, commerce and national security — but only if the nation stays truly committed to the space program.'

NASA's own commitment to the future was exemplified by its work in 1987 toward the return of Space Shuttle flights. August marked the first full-duration test firing of the redesigned Space Shuttle solid rocket motor. The successful firing was the first of several scheduled before the resumption of Shuttle flights once scheduled for February but now planned to begin with mission STS-26 in June 1988.

In establishing the target for launch, Dr Fletcher stated, 'Safely returning the Space Shuttle to flight is NASA's highest priority. Our revised plan for Space Shuttle recovery is ambitious and assumes that we will successfully complete our test and processing objectives. I know I can count on the whole NASA team — and, of course, I include our contractor partners — to move out enthusiastically toward this new goal.'

In October 1987, a new space flight manifest was issued that reflected the 'mixed-fleet' concept — resulting from a major reevaluation of the role of expendable launch vehicles in concert with Shuttle missions.

In late July, NASA completed negotiations with Rockwell International to build the replacement Space Shuttle Orbiter. Using the existing structural spares for construction, OV-105 is scheduled for completion in the April 1991, with a first flight scheduled for early 1992, eight years after *Challenger* was destroyed. The new vehicle will feature the latest upgrades and modifications and will incorporate all new technology evolving from the current return-to-flight activities.

On 8 June, NASA Administrator Dr Fletcher announced a program for elementary and secondary school students to name OV-105, the replacement Shuttle Orbiter. 'It is fitting that students and teachers, who shared in the loss of *Challenger*, share in the creation of its replacement,' he said. The name *Challenger* has been retired, so OV-105 will *not* be named *Challenger II*, as some have suggested.

On 15 September, NASA announced the crew selected for STS-27, a Department of Defense mission targeted for the fall of 1988 aboard the Orbiter *Atlantis*.

Below: **Astronaut George Nelson, in emergency egress training for the much awaited Space Shuttle mission STS-26 in 1988.** ***At right:*** **During STS 26 simulation tests, this NASA flight controller's expression of cautious optimism was a common one.**

PAYLOADS

Late in September, NASA announced the secondary payloads to be carried aboard STS-26. The additional cargo to be flown in *Discovery*'s middeck area was now to include five microgravity experiments; life science, atmospheric science and infrared communications experiments; and two student experiments.

In November, NASA began testing two escape systems that would provide crew egress capability during controlled gliding flight. One escape method uses tractor rockets to extract the astronauts through an open hatch. The other uses a telescoping pole that would extend through the hatch. Crewmembers would escape by sliding down the pole, using a lanyard attached to the rod. After completion of testing and evaluation, a decision will be made on which of the two methods may be incorporated into Shuttle Orbiters.

Meanwhile, in May, Dr Fletcher announced a 'mixed-fleet' plan to acquire expendable launch vehicle (ELV) services for NASA missions to lessen dependence on a single launch system and to preserve the Space Shuttle for missions requiring its unique capabilities. On 22 October, NASA issued its first mixed-fleet manifest for Space Shuttle missions through 1990 and ELVs through 1995.

In launch activities, GOES-H, a NOAA meteorological satellite, was successfully launched by NASA aboard the Delta 179 rocket on 26 February, and the Delta 182 launch vehicle successfully carried an Indonesian Palapa communications satellite, into orbit on 20 March. Both launches took place from Cape Canaveral.

During 1987 NASA also took a major step toward establishment of a permanently manned space station in December with the selection of four aerospace firms to design and build the orbital research base. Boeing Aerospace Company was selected to build the pressurized modules where space station crews will work and live. McDonnell Douglas was chosen to develop the structural framework for the station, as well as most of the major subsystems required to operate the facility.

The international space station complex consists of a manned base, with four pressurized modules capable of supporting a crew of eight people and two unmanned free-flying polar platforms. The manned base will consist of a horizontal truss longer than a football field, three laboratories and a habitat module, four resource nodes, logistics modules and mounting points for attached payloads. Solar arrays will provide 75,000 watts of power to the station, which will be located approximately 300 miles above the earth. NASA's plans call for the space station to be assembled in orbit beginning in 1994, with establishment of permanent manning in 1996.

The space station will provide an unsurpassed research facility for scientific, technology and commercial activities in space in the 1990s and will serve as the cornerstone for continued exploration of the solar system in the next century. The space station will be capable of growth both in size and capability and is intended to operate for several decades, well into the 21st century.

During 1987, the National AeroSpace Plane program basically completed its conceptual design phase. The joint NASA and Department of Defense

NASA continues planning for a space station project in which the Space Shuttle would figure prominently. Artists' conceptions on these pages include a 'dual keel' space station, *below*; a human habitation module that could be used on a manned orbital platform, *at right*; and another space station design, *below right*.

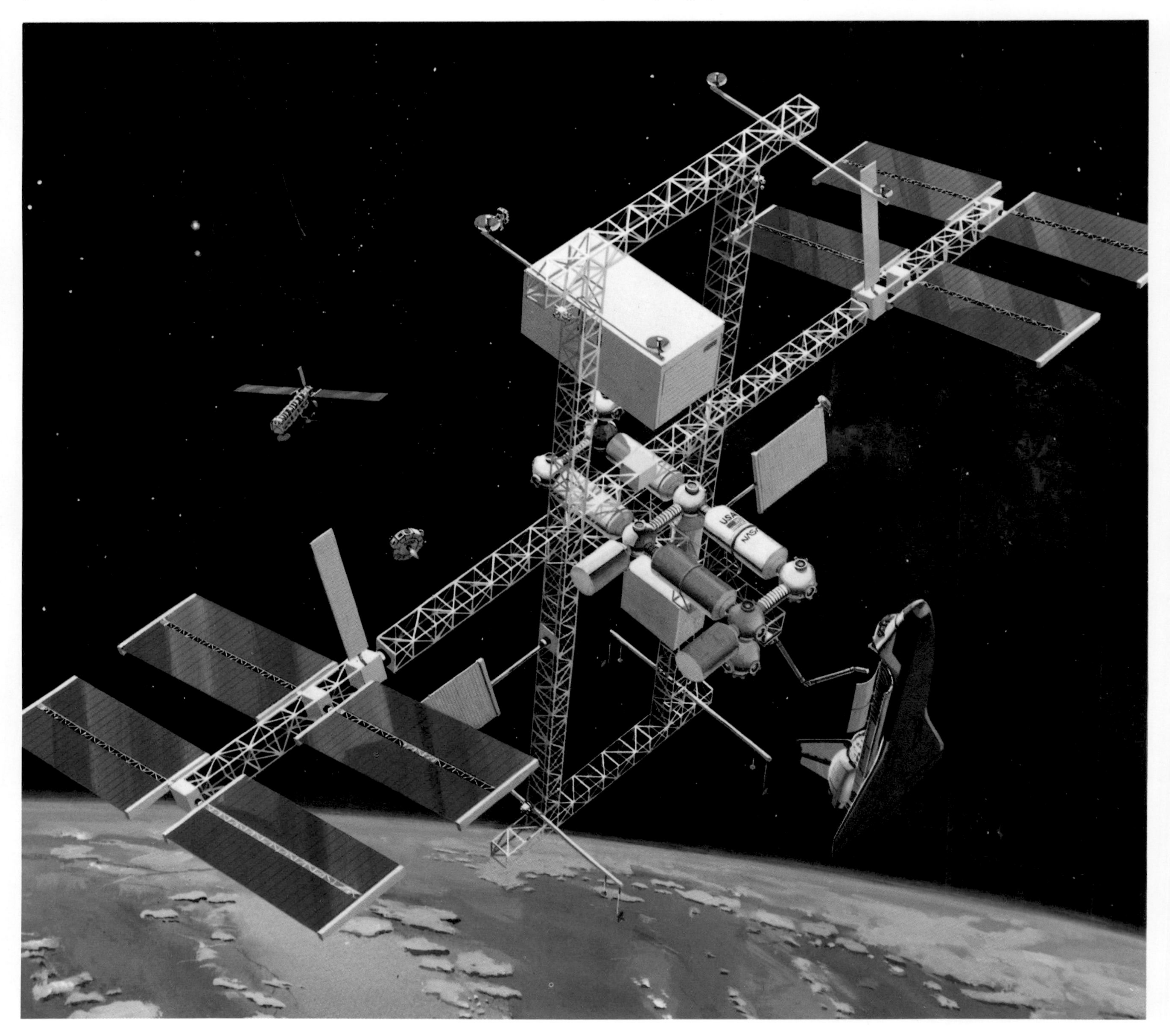

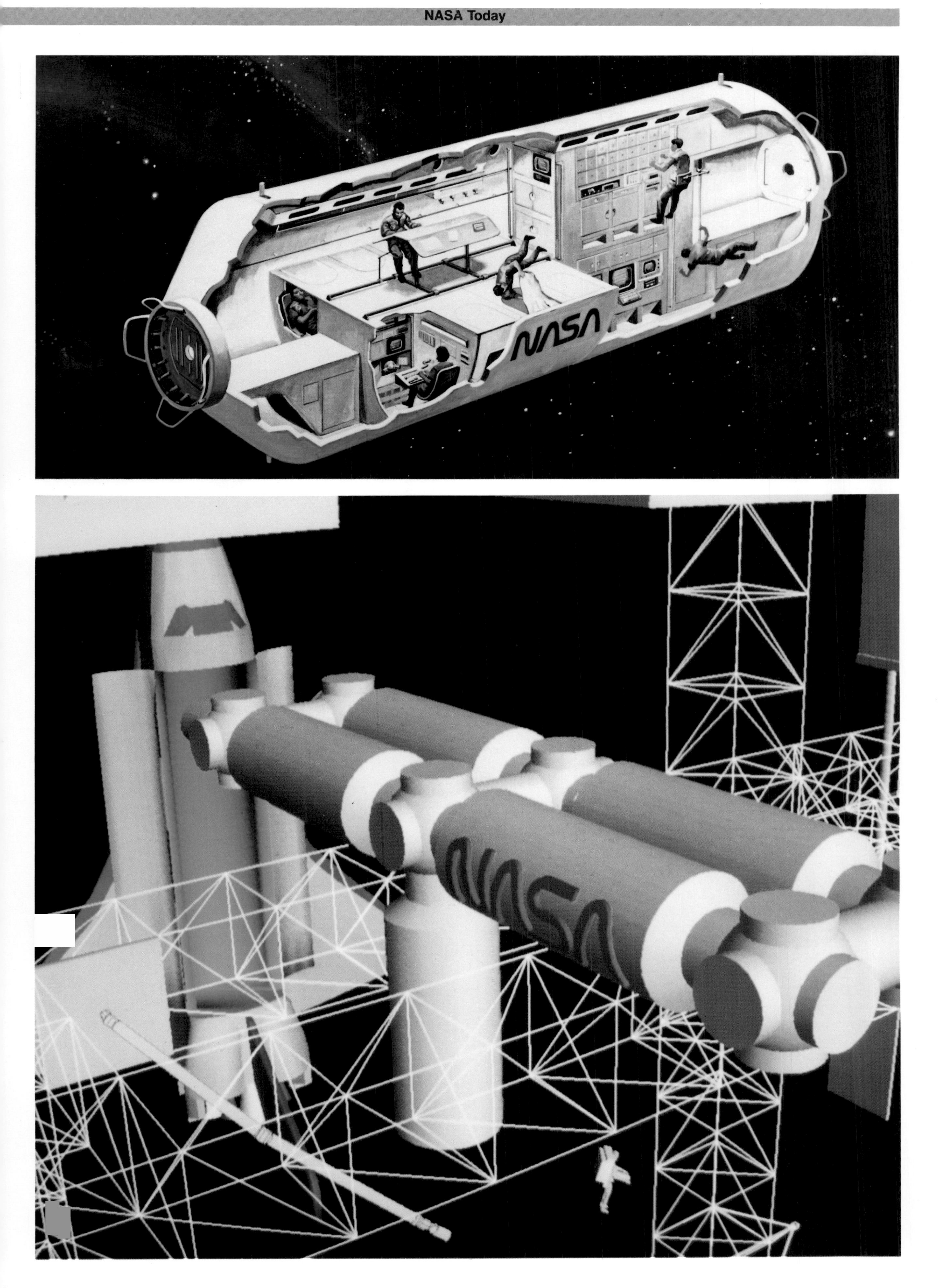
NASA
NASA

program selected three aerospace firms — General Dynamics, McDonnell-Douglas and Rockwell International — to continue vehicle technology development and begin efforts that will result in ground testing of large scale engines and selected aircraft components and in preliminary designs of the X-30 experimental vehicle.

NASA's Deep Space Network began preparing in 1987 for the Voyager spacecraft fly-by of Neptune in August 1989. During 1987, major modifications were under way to increase support capability of the deep space antennas for this and future missions, including Magellan and Galileo. This has been accomplished, in part, by increasing the diameter of the Spanish and Australian 64-meter antennas to 70 meters. The Goldstone antenna will be expanded in 1988.

From August 1986 to August 1987, NASA Astronaut Dr Sally K Ride, who had in 1983 become the first American woman in space, headed a group at NASA Headquarters that studied possible initiatives in pursuit of a new national space objective. The group's final report, 'Leadership and America's Future in Space,' identified four major possibilities for concentrated examination:

- Intensive study of earth systems with the goal of exponentially expanding knowledge required to protect the environment;
- A substantially stepped-up robotic program to explore the planets, moons and other bodies in the solar system;
- Establishment of a scientific base and a permanent human presence on earth's moon; and
- Human exploration of Mars preceded by intensive robotic exploration of the planet.

As part of the report, Ride concluded that the four initiatives identified in her study, along with other potential initiatives, deserved further intensive and systematic consideration to help determine a NASA position on a goal and to follow through after a goal was identified.

On 1 June 1987, NASA Administrator Fletcher announced the creation of the Office of Exploration to coordinate agency activities that would 'expand the human presence beyond earth,' particularly to the moon and Mars.

The office's major responsibilities are to analyze and define missions proposed to achieve the goal of human expansion off the planet, provide central coordination of technical planning studies that involve the entire agency, focus on studies of potential lunar and Mars initiatives, and identify the prerequisite investments in science and advance technology that must be initiated in the near term in order to achieve the initiatives.

During 1987, satellite images of Mexico's Yucatan Peninsula, Central Guatemala and Belize have led to new discoveries about ancient Maya settlement patterns, environmental setting and natural resource use. NASA scientists have found evidence of an ancient river plain, sea level changes and tectonic fault lines, which may have been important geographic elements in shaping the ancient Maya civilization.

Investigators at NASA's Ames Research Center, Mountain View, California, also are using the satellite imagery to detect Maya water sources, such as natural wells and ponds, and compare their locations to those of ancient Maya ruins. Investigators believe the remote sensing project will help resolve a central question in Maya studies — how the Maya built a sophisticated civilization in a relatively resource-poor environment? They also hope to understand the mysterious cycles of expansion and decline that characterized Maya civilization. Many scholars believe that environmental problems, including misuse of resources, may have led to the periods of decline. The Maya civilization, noted for elaborate temples, advanced mathematics and astronomy and large-scale agriculture, spread across Central America from 2000 BC until the Spanish conquest in the 16th Century.

Using Earth Observation Satellite and Earth Resources Observation Systems imagery from the Landsat 5 satellite, which has an advanced multi-spectral sensor called Thematic Mapper, the NASA-Ames researchers have imaged more than 24,000 square miles of the northern Yucatan Peninsula. A total of 50,000 square miles were imaged

NASA's Final Countdown to the Return of Manned Spaceflight

In March 1988, NASA named crew members for STS-29, STS-30 and STS-31, the three Spaceshuttle missions scheduled to fly in January, April and June 1989. This announcement concludes crew selection activities planned prior to the resumption of Shuttle flights in August of this year. STS-31 will feature deployment of the Hubble Space Telescope originally

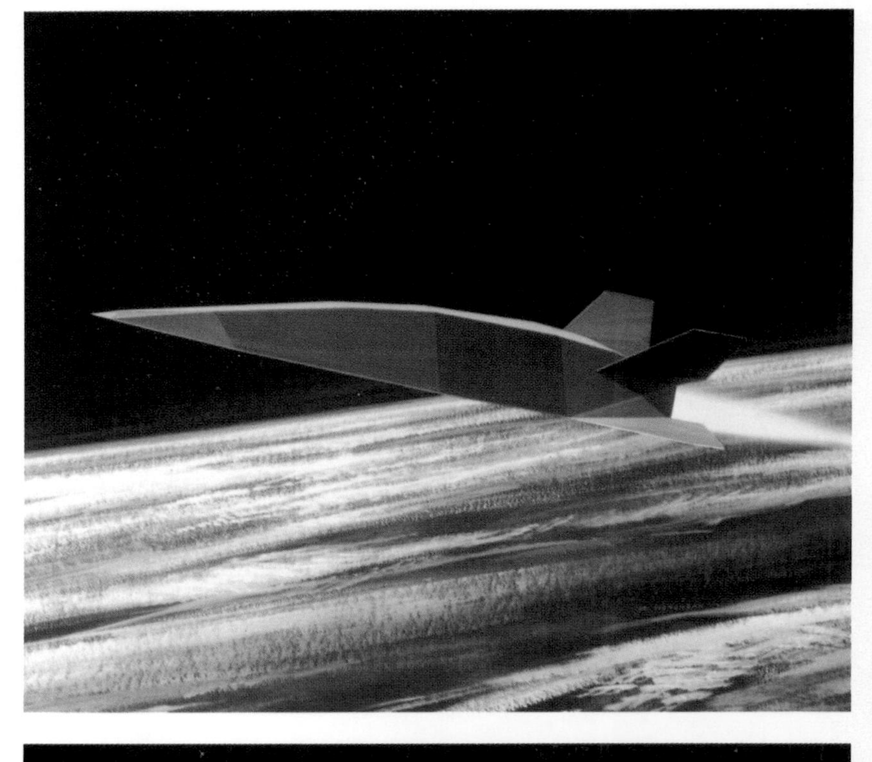

Artist's views of three National Aerospace Plane concepts — *at top,* the McDonnell Douglas Transatmospheric Vehicle ('TAV,' preliminary design); *above middle,* the Lockheed TAV; and *above,* the Rockwell NASP (X-30). *At right: Discovery* STS-26 breaks the Space Shuttle program's 2.6-year grounding on 29 September 1988.

USA
NASA

planned for June 1986, and now targeted for launch 1 June 1989. No date for STS-28 was given in the March 1988 announcement, although it was postponed until *after* STS-29.

A decision to interchange the STS-29 and STS-28 missions was seen by NASA to ease the orbiter processing flow and enables NASA to maintain the required launch windows for two interplanetary missions in 1989—Magellan, a mission to map the planet Venus in April, and Galileo, a cooperative project with Germany to survey Jupiter and its moons, in October. The Hubble Space Telescope also maintains its flight assignment date of June 1989.

Astro-1, a Spacelab mission designed to study the universe in the ultraviolet spectrum is being reconfigured to enhance the study of Supernova 1987A, an event that has drawn the attention of astrophysicists from around the world. The Broad-Band X-Ray Telescope has been added to complement the Astro-1 mission now slated to fly on STS-35 in November 1989.

Taking advantage of the recently-announced Shuttle downweight additional capability, Spacelab missions are now planned to fly aboard the Orbiter *Columbia* (OV-102), which was not previously possible. Two Spacelab payloads have been assigned flights in 1990—a Spacelab Life Sciences mission in March and the first of the Atmospheric Laboratory for Applications and Science mission series, ATLAS-1, scheduled for September. The Gamma Ray Observatory, moved forward in the projected schedule, is now slated for March 1990 and the Ulysses mission to study the sun and its environment, remains in its projected October 1990 launch date.

Another important addition to the manifest is a mission to retrieve the Long Duration Exposure Facility (LDEF) in July 1989. Launched by the Space Shuttle in April 1984, LDEF originally was scheduled for retrieval in March 1985. The LDEF retrieval mission will replace Astro-1 as the payload for STS-32.

The *Discovery* STS-26 crew wore vacation apparel for this photograph *(below)*. They are, from left to right front, John Lounge, David Hilmers, mission commander Frederick Hauck, George Nelson and Shuttle pilot Richard Covey. *At right:* An STS-26 photo of Discovery's tailfin, lit up by the earth/sun corona. STS-26 launched the Tracking and Data Relay Satellite/Inertial Upper Stage, at *far right*.

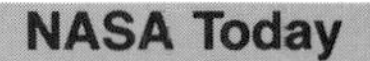

Below: Discovery STS-26 comes in for a landing at NASA's Dryden Flight Research Center (at Edwards AFB in California), on 3 October 1988. *At right* is the scene at Mission Control, in the Johnson Space Center in Houston, Texas, as *Discovery's* wheels first touch down on the dry lake bed at Dryden. *At far right,* the crew of *Discovery* STS-26 (see also the caption on page 180) poses for a group portrait with then-Vice President George Bush, shortly after *Discovery's* triumphant four-day mission.

Despite a delay caused by unexpected upper atmospheric winds, the *Discovery* STS-26 launch went well. Decision to launch this mission was made by Deputy Director of NASA Space Flight Operations Robert Crippen—himself a veteran of four Space Shuttle missions, including the very first, STS-1, which he flew with John Young—and was governed by approximately 2500 launch commit criteria, or about one-fifth more than were used for previous Shuttle flights. The Orbiter *Discovery* incorporates 210 changes from pre-*Challenger* Orbiter configurations, and the Shuttle Transportation System's Solid Rocket Boosters have been completely redesigned. Approximately 35 modifications have been made to the Orbiter's main engines, and 100 changes have been made to the Orbiter's software. In addition, for its ascent, *Discovery* flew a trajectory that would allow the crew to fly the Shuttle to a landing site in the event of engine failure. The Tracking and Data Relay Satellite launched by STS-26 will expand US space-based relay coverage to 85 percent of the earth, given an equatorial orbit.

Discovery

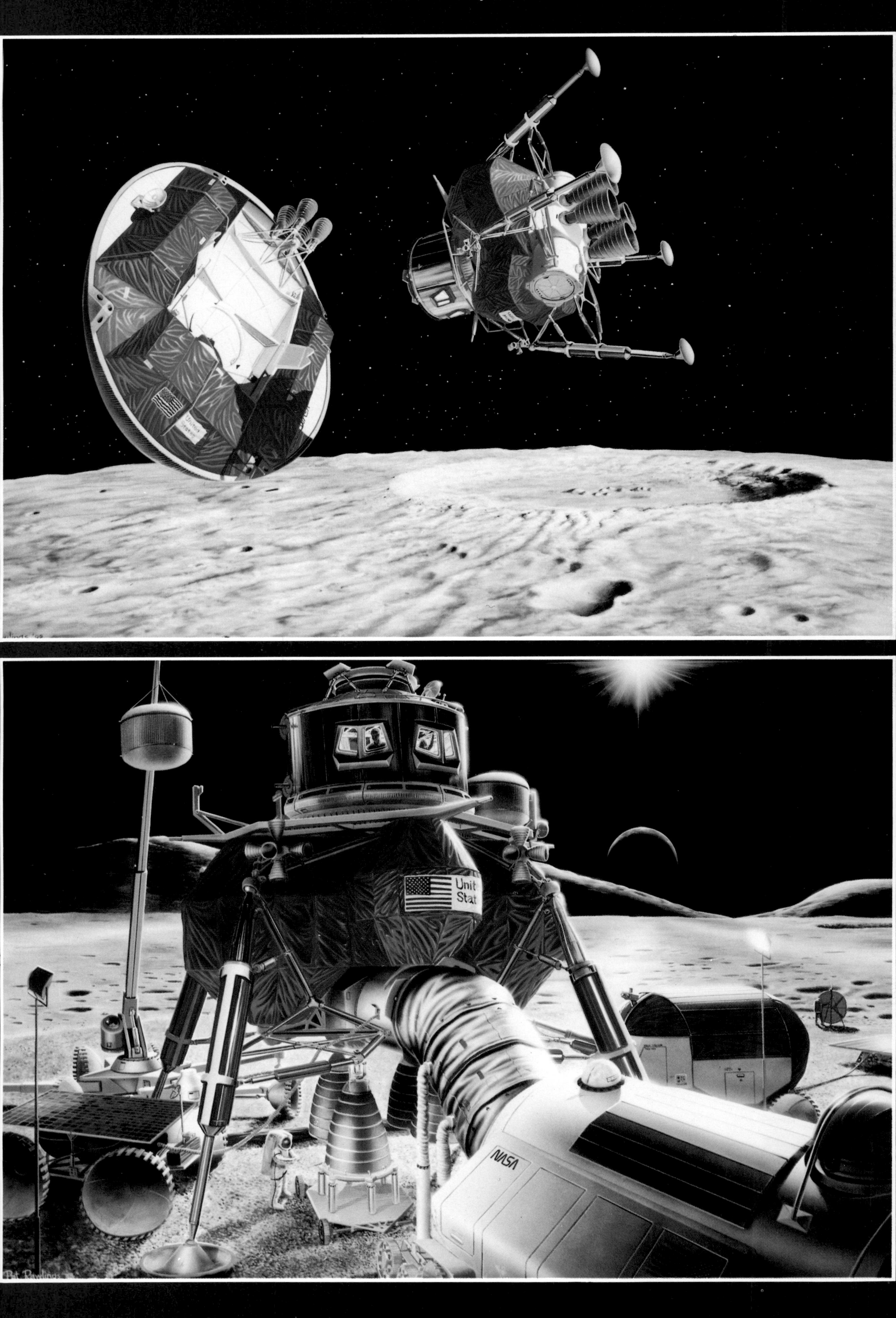
United States
Unit
Stat
NASA

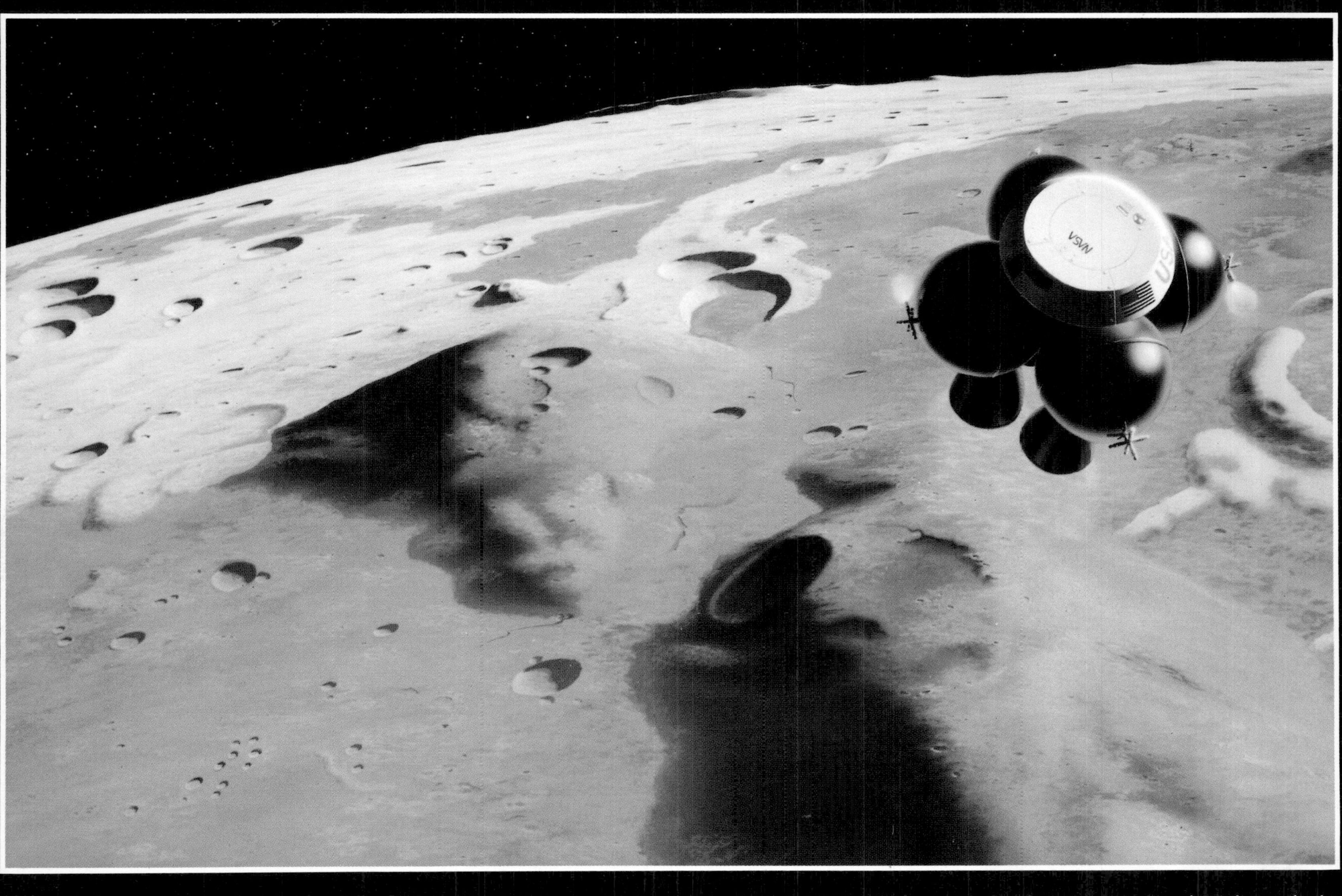

Above left: A conceptual drawing of a lunar Orbiter/Lander Vehicle that may one day result from the NASA-sponsored Lunar Base Systems Study. The saucer-shaped Lander can be modularly altered to handle personnel or cargo, or a combination of both, and is here heading for the lunar surface. *Below left:* This artwork, based on the same study, portrays a lunar landing facility, with a just-landed Descent Module at rear, and a lunar transport vehicle—ready to take passengers to a lunar base—in the foreground. *Above:* An artist's conception of a futuristic lunar sample return robot 'bus.' *Below:* Lunar Base Systems Study artwork depicting a hypothetical oxygen-producing plant on earth's moon. This plant would use indigenous lunar materials to produce oxygen by a chemical reaction process.

Looking Toward the Future

During 1987 and 1988, two National Aeronautics and Space Administration centers undertook studies of concepts for designing and building a robot space vehicle that would travel from earth to Mars, land and explore the martian surface and return an 11-pound sample of rock and soil to earth. Human exploration of Mars, preceded by robotic missions including the sample return, is one of four space initiatives under study by NASA to help the nation determine its next major goal in space. If approved and funded, the sample return mission would be launched in 1998 and return with martian soil, rock and atmosphere samples in 2001. NASA allocated $1.2 million for studies in Fiscal Years 1987–88.

Task teams at NASA's Johnson Space Center (JSC), Houston, and at NASA's Jet Propulsion Laboratory (JPL), Pasadena, California, are outlining preliminary scenarios on how a Mars sample return mission could be flown. JPL would study a Mars Orbiter spacecraft and an automated surface rover equipped with sample gathering scoops and tongs. The Orbiter and Mars surface craft would depend heavily on preprogrammed computer software to manage tasks, since radio commands take almost an hour to reach Mars from earth. JPL will have overall project management. JPL managed the Viking Mars lander missions in 1975–76, the Voyager planetary missions and most other United States unmanned planetary exploration projects.

As now envisioned, JSC would study a spacecraft capable of entering and slowing down in the martian atmosphere — estimated at one percent the density of the earth's atmosphere — to land gently on the planet's surface. JSC also would study the vehicle stage for lifting off Mars to rendezvous and dock with an orbiting spacecraft prior to return to earth with the surface samples.

Returned martian soil, rock and atmosphere samples would undergo laboratory analysis, much like samples from earth's moon during the Apollo lunar landing program. The space station probably will be used for initial study and analysis of the returned samples.

NASA Administrator Fletcher said on 14 April 1988 that the moon, rather than Mars, may be the best initial destination for possible joint US/USSR manned missions. 'Going to the moon together would give the two leading spacefaring nations in the world an opportunity to build a stable base for further cooperation, which could, one day, lead to a cooperative mission to Mars,' he said.

'Flying out to Mars together before building such a foundation could, for several reasons, be less practical,' Dr Fletcher told participants at the National Space Symposium in Colorado Springs, Colorado.

Below: **An illustration of a Mars sample return mission that is as yet only being thought of by NASA.** ***At below left:*** **A variant artist's concept of a Mars sample return mission.** ***At left*****: An artist's concept of the Mars Observer Mission planetary and atmospheric probe, scheduled for launch from the Space Shuttle in 1992.**

NASA

APPENDIX: MANNED MISSIONS AND DURATIONS

Mercury Program

	Crew	*Date*
Mercury 3	Shepard	5 May 1961 (00:15:22)
Mercury 4	Grissom	21 July 1961 (00:15:37)
Mercury 6	Glenn	20 February 1962 (04:55:23)
Mercury 7	Carpenter	24 May 1962 (04:56:05)
Mercury 8	Schirra	3 October 1962 (09:13:11)
Mercury 9	Cooper	15–16 May 1963 (34:19:49)
Total time:	Mercury Program	(53:55:27)

Gemini Program

	Crew	*Date*
Gemini 3	Grissom, Young	23 March 1965 (04:53:00)
Gemini 4	McDivitt, White	3–7 June 1965 (97:56:11)
Gemini 5	Cooper, Conrad	21–29 August 1965 (190:55:14)
Gemini 7	Borman, Lovell	4–18 December 1965 (330:35:31)
Gemini 6	Schirra, Stafford	15–16 December 1965 (25:51:24)
Gemini 8	Armstrong, Scott	16 March 1966 (10:41:26)
Gemini 9	Stafford, Cernan	3–6 June 1965 (72:21:00)
Gemini 10	Young, Collins	18–21 July 1966 (70:46:39)
Gemini 11	Conrad, Gordon	12–15 September 1966 (71:17:08)
Gemini 12	Lovell, Aldrin	11–15 November 1966 (94:34:31)
Total time:	Gemini Program	(969:52:04)

Apollo Program

	Crew	*Date*
Apollo 7	Schirra, Eisele, Cunningham	11–22 October 1968 (260:09:03)
Apollo 8	Borman, Lovell, Anders	21–27 December 1968 (147:00:42)
Apollo 9	McDivitt, Scott, Schweickart	3–13 March 1969 (241:00:54)
Apollo 10	Stafford, Young, Cernan	18–26 May 1969 (192:03:23)
Apollo 11	Armstrong, Collins, Aldrin	16–24 July 1969 (195:18:35)
Apollo 12	Conrad, Gordon, Bean	14–24 November 1969 (244:36:25)
Apollo 13	Lovell, Swigert, Haise	11–17 April 1970 (142:54:41)
Apollo 14	Shepard, Roosa, Mitchell	31 January–9 February 1971 (216:01:57)
Apollo 15	Scott, Worden, Irwin	26 July–7 August 1971 (295:11:53)
Apollo 16	Young, Mattingly, Duke	16–27 April 1972 (265:51:05)
Apollo 17	Cernan, Evans, Schmitt	7–19 December 1972 (301:51:59)
Total time:	Apollo Program	(2719:29:00)

Skylab Program

	Crew	Date
Skylab 2	Conrad, Kerwin, Weitz	25 May – 22 June 1973 (672:49:49)
Skylab 3	Bean, Garriott, Lousma	28 July – 25 September 1973 (1427:09:04)
Skylab 4	Carr, Gibson, Pogue	16 November 1973 – 8 February 1974 (2017:15:32)
Total time:	Skylab Program	(4117:14:25)

Apollo-Soyuz Program

	Crew	Date
Test Project	Stafford, Brand, Slayton	15 – 24 July 1975 (217:28:23)

Space Shuttle Program

	Crew	Date
STS-1	Young, Crippen	12 – 14 April 1981 (54:20:32)
STS-2	Engle, Truly	12 – 14 November 1981 (54:13:13)
STS-3	Lousma, Fullerton	22 – 30 March 1982 (192:04:45)
STS-4	Mattingly, Hartsfield	27 June – 4 July 1982 (169:09:40)
STS-5	Brand, Overmyer, Allen, Lenoir	11 – 16 November 1982 (122:14:26)
STS-6	Weitz, Bobko, Peterson, Musgrave	4 – 9 April 1983 (120:23:42)
STS-7	Crippen, Hauck, Ride, Fabian, Thagard	18 – 24 June 1983 (146:23:59)
STS-8	Truly, Brandenstein, D Gardner, Bluford, W Thornton	30 August – 5 September 1983 (145:08:43)
STS-9	Young, Shaw, Garriott, Parker, Lichtenberg*, Merbold*	28 November – 8 December 1983 (247:47:24)
41-B	Brand, Gibson, McCandless, McNair, Stewart	3 – 11 February 1984 (191:15:55)
41-C	Crippen, Scobee, van Hoften, Nelson, Hart	6 – 13 April 1984 (167:40:07)
41-D	Hartsfield, Coats, Resnik, Hawley, Mullane, C Walker*	30 August – 5 September 1984 (144:56:04)
41-G	Crippen, McBride, Ride, Sullivan, Leestma, Garneau*, Scully-Power*	5 – 13 October 1984 (197:23:37)
51-A	Hauck, D Walker, D Gardner, A Fisher, Allen	8 – 16 November 1984 (191:44:56)
51-C	Mattingly, Shriver, Onizuka, Buchli, Payton*	24 – 27 January 1985 (73:33:27)
51-D	Bobko, Williams, Seddon, Hoffman, Griggs, C Walker,* Garn*	12 – 19 April 1985 (167:54:00)
51-B	Overmyer, Gregory, Lind, Thagard, W Thornton, van den Berg*, Wang*	29 April – 6 May 1985 (168:08:47)
51-G	Brandenstein, Creighton, Lucid, Fabian, Nagel, Baudry*, Al-Saud*	17 – 24 June 1985 (169:39:00)
51-F	Fullerton, Bridges, Musgrave, England, Henize, Acton,* Bartoe*	29 July – 6 August 1985 (190:45:26)
51-I	Engle, Covey, van Hoften, Lounge, W Fisher	27 August – 3 September 1985 (170:27:42)
51-J	Bobko, Grabe, Hilmers, Stewart, Pailes*	3 – 7 October 1985 (97:14:38)
61-A	Hartsfield, Nagel, Buchli, Bluford, Dunbar, Furrer,* Messerschmid,* Ockels*	30 October – 6 November 1985 (168:44:51)
61-B	Shaw, O'Connor, Cleave, Spring, Ross, Neri-Vela,* C Walker*	26 November – 3 December 1985 (165:04:49)
61-C	Gibson, Bolden, Chang-Diaz, Hawley, G Nelson, Cenker*, B Nelson*	12 – 18 January 1986 (146:03:51)
51-L	Scobee, Smith, Resnik, Onizuka, McNair, Jarvis,* McAuliffe*	28 January 1986 (0:01:13)
STS-26	Hauck, Covey, Hilmers, Nelson, Lounge	29 September 1988 (96:00:09)
STS-27	Gibson, Gardner, Mullany, Shepherd, Ross	2 December 1988 (105:06:19)

*(Payload specialists, not astronauts)

INDEX

Below: **Technicians at Rockwell International's Palmdale facility mate a right wing assembly to the fuselage of OV-105, the new Space Shuttle that is meant to replace OV-99 *Challenger*.**

Overleaf: **An Apollo astronaut stands on the moon, with the earth above and the American flag hanging as it could only in the windless, low gravity environment of the lunar surface. This scene was born of a longing in the human spirit that has burned since the first explorers put a log in a river and hopped on for the ride — and the saga would seem to be far from finished.**